Hitler's Engineers

The Hidden Story Of Nazi Technology In America

GEW Humanities Group

Global East-West (London)

Copyright © 2025 by GEW Humanities Group.

" History For All" collection, edited by Hichem Karoui.

All rights reserved.

No portion of this book may be reproduced in any form without written permission from the publisher or author, except as permitted by copyright law.

Contents

Introduction   1
Unveiling the Legacy of War

1. The Rise of Fascism   15
   A World in Turmoil

2. Operation Paperclip   33
   The Recruitment of Genius

3. The Shadowy Collaboration   49

4. Rockets And The Space Race   65

5. Triumphs and Trials   81
   The Wonders of Nazi Engineering

6. Jet Propulsion   99
   Redefining Warfare And Aviation

7. The Pursuit Of The Allies   115

8. Medicine and Morality   133
   The Ethical Dilemma

9. The Secrets Of Special Operations   151

10. The Enigma of Cryptography   167
    Codes and Secrets

11.	The Invisible Hand Nazi technology and the Cold War	185
12.	The Mysteries Of Area 51 Fact or Fiction?	203
13.	Veiled Innovations Secret Projects	217
14.	The Industrial Powerhouse Techniques and transformation	231
15.	Spy Games	247
16.	Impact On Modern Industry And Technology	263
17.	The Legacy of Resistance	281
18.	Reflections Ethical Implications and the Future of Innovation	299

References 317
For Further Reading

Introduction
Unveiling the Legacy of War

Contextualising war

The geopolitical landscape that preceded the major conflicts explored in this narrative is integral to understanding the multiple dimensions of war. To put war into context, we must embark on a historical odyssey through the complex web of international relations, power dynamics and ideological tensions that spanned continents. The interplay of territorial ambitions, ideological fervour and imperialist designs at the beginning of the 20th century paved the way for a cataclysmic global clash. The smouldering embers of unresolved grievances from previous conflicts, combined with the emergence of aggressive expansionist policies, created a volatile powder keg just waiting to be lit. The shifting alliances and diplomatic manoeuvring of the great powers intertwined with regional disputes, fuelling the conflagration that engulfed the world.

The economic upheavals and stark disparities resulting from the First World War created fertile ground for the seeds of discontent and disillusionment, nurturing radical ideologies that sought to redraw the geopolitical map. Furthermore, the Great Depression cast a lasting shadow over nations, creating conditions conducive to the rise of populist demagogues and extremist movements. These broader societal upheavals reverberated through the corridors of power, influencing foreign policy decisions and triggering conflicts that would alter the course of history. Thus, the rich tapestry of intertwining events and motivations illuminates the complex geopolitical cauldron that brewed the storm of global conflict. It is against this backdrop of conflicting interests and ideologies that the stories in the following chapters unfold, describing the profound impact of intersecting geopolitical currents and the indelible scars they left on civilisation.

The moral landscape of conflict

In the aftermath of war, alongside the physical devastation, there remains a profound moral impact that reverberates across generations. The ethical complexities inherent in armed conflict have shaped and challenged societies and individuals, leaving an indelible mark on the collective consciousness. Examining the moral landscape of conflict requires a delicate balance between historical perspective and contemporary reflection, delving into the multiple dimensions of righteousness and responsibility. At the heart of this exploration lies the fundamental question of justifying actions taken in wartime. The parameters of morality are tested as conflicts unfold, forcing society to confront its own ethical compass and grapple with the implications of choices made in the crucible of adversity. The dichotomy between the necessity of certain actions in wartime and their alignment with ethical principles highlights the complexity of navigating the moral terrain in the heat of the moment. Furthermore, the moral fabric of conflict extends beyond simple decisions made on the battlefield, permeating international relations, political strategies, and the treatment of prisoners and civilians. It encompasses the formidable responsibility of leaders to uphold moral standards amid the chaos of war and to strike a precarious balance between military demands and human rights. Exploring the complexities of moral decision-making during conflict requires an examination not only of overt acts, but also of the more subtle violations against humanity that persist in the shadows of war—from propaganda and psychological warfare to the deliberate targeting of cultural heritage.

The moral landscape of conflict is fraught with lasting ethical ramifications that extend far beyond the cessation of hostilities. The echoes of war's moral challenges resonate in post-war justice, reparations, and ongoing efforts at healing and reconciliation. These enduring moral dilemmas demand attention as they shape the present and define the collective identity of nations, testifying to the interconnectedness of ethical responsibility and the aftermath of war. Confronting the moral legacy of conflict requires not only looking back at the past, but also critically assessing the enduring lessons learned from these ethical crucibles. This introspection sheds light on the complex interaction between moral imperatives and the demands of war, serving as both a cautionary tale and a catalyst for greater understanding, empathy, and unwavering dedication to preserving ethical values in the face of adversity.

Technological advances born of combat

In the crucible of conflict, humanity has often witnessed the birth of remarkable technological advances. Historically, wars have served as catalysts for rapid innovation and progress in various fields, fuelling advances in weaponry, communication, transport, medicine and many other areas. From the development of radar systems and nuclear capabilities to the evolution of aviation and computer technology, the crucible of combat has propelled societies into uncharted territory in terms of scientific and engineering prowess. Great minds, often under the constraints of wartime demands, rose to the challenges posed by adversaries, resulting in revolutionary inventions that transformed the battlefield and left an indelible mark on the course of human civilisation. The urgency of war prompted nations to invest heavily in research and development, leading to the creation of cutting-edge tools and methodologies that have profound implications beyond the immediate theatre of war. It is imperative to recognise these technological advances born of combat and their lasting impact on the fabric of our modern world. While the ethical dimensions of these innovations require examination and contemplation, it is undeniable that many of these breakthroughs have pushed the boundaries of human achievement. We aim to examine some of the most crucial and significant technological advances made in the chaos of war, highlighting the transformative power of adversity and the legacy of ingenuity that continues to resonate throughout the annals of history.

Personal and collective memories

War has a profound impact on individuals and societies, leaving behind lasting memories that shape cultural and historical narratives. The personal memories of those directly involved in conflict provide a fascinating insight into the human experience of war. These memories offer a nuanced understanding of the emotional, psychological, and physical toll imposed by war. Beyond the individual, collective memories encapsulate the broader social and cultural repercussions of war. These memories are imbued with the shared experiences of communities, nations, and even global alliances, and bear witness to humanity's endurance and resilience in the face of adversity. Personal and collective memories are an integral part of the post-war landscape, influencing ideologies and shaping future generations' perceptions of conflict. These memories, passed down through oral traditions, literature, art, and commemorative events, become woven into the fabric of our societal consciousness. Preserving personal and collective memories ensures that the lessons of war are not lost to time, compelling us to contemplate the tragic consequences and strive to build a peaceful and prosperous future. Through the prism of memory, we confront the complexities of war, recognising its profound impact while encouraging a collective commitment to learning from history and fostering international harmony.

The Impact of War on Political Borders and Power

The effect of war on political borders and power is a multifaceted phenomenon that resonates deeply in the fabric of world history. When nations engage in conflict, political boundaries are often redefined and reshaped. In the aftermath of wars, territorial adjustments, annexations, and even the creation of entirely new geopolitical entities frequently occur. These changes not only reflect shifting power dynamics, but also serve as a lasting testament to the impact of war on the geopolitical landscape. These changes can have far-reaching implications, influencing diplomatic relations, resource allocation, and regional stability for generations to come.

The exercise of power within and between states undergoes significant transformations in times of war. The struggle for dominance and control of territories intensifies, leading to shifts in alliances, the emergence of new coalitions, and the formation of power blocs. The complex interaction of socio-political forces during and after war often creates power vacuums, ideological divisions, and opportunities for strategic realignment that reverberate across continents. With the rise of military power and the pursuit of strategic objectives, political powers manoeuvre to assert their influence over key regions, exploit resources, and secure advantageous positions over their adversaries. This relentless quest for power serves as a catalyst for geopolitical tensions and territorial rivalries, shaping the trajectories of international relations and reconfiguring the global balance of power. Simultaneously, the impact of war on political boundaries extends beyond overt territorial changes to encompass the complex interweaving of identity, culture, and sovereignty. Ethnic and national identities are often redefined or reinforced in the crucible of conflict, with communities rallying around shared narratives of resilience and sacrifice.

The reconfiguration of political boundaries can give rise to profound feelings of displacement, belonging and historical injustice, manifesting as lasting social and cultural divisions that reverberate through the pages of history. Ultimately, studying the impact of war on political boundaries and power provides invaluable insights into the complex tapestry of global politics and the evolution of the art of governance. By examining the link between conflict, power dynamics, and territorial transformations, we gain a deeper understanding of the lasting legacy of war and the indelible marks it leaves on the contours of geopolitical power.

Economic Shockwaves: From National to Global

The consequences of war are often felt far beyond the battlefield, extending to all aspects of society, including the complex network of economic structures that support nations. The immense toll of global conflicts reverberates through economies, sending shockwaves from national borders to the interconnected realm of the global economy. Post-war reconstruction efforts and industrial conversion are crucial moments that redefine the economic landscape. While some regions struggle with the burden of rebuilding, others witness abrupt changes in their economic fortunes as resources and capital flow in new directions. This reconfiguration is often a source of conflict and opportunity, redrawing traditional trade routes and creating new alliances. The transformation of national industries is a direct consequence of war, with some sectors booming while others decline.

In their quest for recovery, nations are strategically investing in technology, infrastructure, and human capital, with the intention of emerging stronger and more resilient. The mobilisation of resources on such a scale paves the way for international cooperation and competition, leading to changes in the global economic hierarchy. As various societies adapt to the challenges of the post-war period, economic policies evolve, aiming to stabilise markets and promote growth.

At the heart of this tumultuous period lies a complex interaction between national economies and the emerging dynamics of globalisation, ultimately paving the way for a new era of economic interdependence. Faced with the reality of unprecedented societal change, policymakers and business leaders sought to harness the potential of global economic integration, discerning opportunities amid the upheaval. The effects of the war transcended borders, compelling nations to coordinate efforts to rebuild destroyed economies, forging ties that extended beyond traditional trade partnerships.

Simultaneously, economic upheaval poses myriad challenges, from managing inflation and unemployment to recalibrating financial systems. From one continent to another, the echoes of war resonate in fiscal policies, trade and investment strategies, shaping the trajectory of prosperity and influencing the contours of emerging markets. It is within this multifaceted framework that profound shifts in global economic power are taking root, giving rise to a complex tapestry of opportunities and constraints that will define the economic landscape for generations to come.

Cultural shifts and ideological echoes

The upheavals caused by global conflicts led to a profound transformation of cultural dynamics and created lasting ideological echoes that continue to resonate today. Through the crucible of war, societal paradigms shifted as perceptions of identity, authority and national allegiance were redefined.

Cultural expressions evolved in response to prevailing tensions, influencing art, literature, music, and cinema. These changes reflected the collective struggle for meaning and belonging in a world fractured by ideological divisions. The rise of totalitarian regimes imposed a homogenised cultural framework, often suppressing diverse expressions of creativity and individuality. Yet within these regimes, underground movements emerged, clandestinely preserving cultural heritage and encouraging subversive forms of artistic resistance. The legacy of such defiance illustrates the resilience of the human spirit in the face of oppressive ideology.

As borders were redrawn and populations displaced, the blending of diverse cultures spawned unprecedented innovation and adaptation. Social customs, traditions, and languages intermingled, giving rise to new syncretic expressions and cultural identities. This fusion of sensibilities catalysed the evolution of global interconnectedness, enriching the tapestry of humanity with a multiplicity of narratives and perspectives. The evolution of cultural ethos during this tumultuous period also facilitated the spread of ideological paradigms, both benevolent and malevolent, creating lasting legacies that continue to shape contemporary discourse. The vestiges of propaganda and indoctrination persist as cautionary tales, prompting societies to critically evaluate the ethical ramifications of unchecked political influence over cultural production and dissemination. In exploring these transformations, it becomes evident that the clash of ideologies during wartime not only precipitated cultural realignment, but also embedded lasting repercussions in the fabric of societal consciousness. Thus, examining these cultural shifts offers invaluable insights into the resilience of human creativity and the complex interplay between ideology, power, and the construction of historical narrative.

Documenting History: Sources and Challenges

When delving into the annals of history, one must navigate a complex web of sources and challenges that arise in the pursuit of truth and accuracy. The documentation of historical events relies on a myriad of resources, including primary sources such as official documents, testimonies, and artefacts, as well as secondary sources such as academic works and analyses. Each source has its own intrinsic value, offering unique perspectives that collectively contribute to the mosaic of historical understanding. However, this abundance of sources poses the problem of discerning authenticity, as historical accounts are often marred by bias, misinformation, and conflicting interpretations. As historians, we are faced with the task of meticulously sifting through these sources, critically evaluating their reliability and context, and ultimately reconstructing a narrative based on evidence and objectivity.

The nature of historical documentation extends beyond textual records to encompass visual materials, oral histories, and archaeological finds, each adding layers of complexity and richness to the narration of the past. While these sources offer invaluable insights, they also present challenges in terms of preservation, interpretation, and accessibility. Preservation involves protecting fragile historical materials from the erosive forces of time and the environment, requiring meticulous archival practices and conservation efforts to ensure a lasting legacy. Interpretation is central to historical study, as historians must decipher and analyse the nuances contained within sources to construct an accurate and comprehensive representation of events.

Accessibility is an ongoing challenge, as historical documents may be scattered across various locations, restricted by institutional barriers, or hidden in the labyrinthine depths of archives. Overcoming these obstacles requires collaboration, technological innovation, and ethical stewardship to make history accessible to all who seek to engage with it. Thus, the pursuit of documenting history is not simply a scholarly endeavour, but a profound responsibility: to honour the past, strengthen the present, and illuminate the future through the unwavering pursuit of truth.

Overview of the themes explored in the book

In the following chapters, we will examine a myriad of interconnected themes that emerged from the legacy of war:

The Rise of Fascism will transport us to a time of ideological upheaval, where we will explore the origins and influence of the political movements that reshaped the global landscape. *Operation Paperclip* will allow us to shed light on the recruitment of brilliant minds, highlighting the ethical issues faced by nations seeking intellectual prowess in the aftermath of conflict. Our exploration of *Shadow Collaboration* will uncover the clandestine alliances and complex network of interactions that shaped the post-war period. *Rockets and the Space Race* will propel us into a world of scientific wonders and geopolitical aspirations, revealing the duality of innovation born of adversity. *Triumphs and Trials: The Wonders of Nazi Engineering* will confront us with the moral ambiguities of technological achievements intertwined with tyranny. *Jet Propulsion* will introduce us to the rapid evolution of the aviation industry, alongside the redefinition of warfare capabilities. As we discover *The Pursuit of Allies*, a nuanced understanding of pursuit and preservation emerges, highlighting the desire to possess or suppress revolutionary knowledge. *Medicine and Morality* will highlight the ethical dilemmas arising from advances in wartime healthcare and the unintended consequences of progress. *The Secrets of Special Operations* reveals the clandestine efforts and intelligence strategies employed to gain military advantage. *The Enigma of Cryptography* will lift the veil on the mysteries of codes and secrecy, highlighting their essential role in shaping global conflicts and power dynamics. *The Invisible Hand* delves into the tangled web of technology, politics, and the Cold War—a story of influence, innovation, and espionage. *The Mysteries of Area 51* will navigate between fact and folklore, addressing the enigmatic narratives that have captured the public imagination. In *Veiled Innovations*, we discover the clandestine projects that have left an indelible mark on modern civilisation, shrouded in extraordinary secrecy and intrigue. Industrial Power will highlight the techniques of transformation and adaptation implemented by industries during wartime, which have had a profound influence on contemporary technology and infrastructure. *Spy Games* will take us into a labyrinth of intrigue and subterfuge, offering a glimpse into the shadowy world of espionage and covert operations at the site. *Impact on Modern Industry and Technology* will chronicle the lasting impact of wartime innovations on contemporary industrial and technological landscapes, highlighting the enduring DNA of conflict within our progress. *The Legacy of Resistance* will highlight stories of defiance and resilience, weaving a tapestry of individuals, communities and movements that weathered the storm of war. Finally, *Reflections* will pave the way for contemplation of the ethical implications arising from the converging forces of innovation, war and the future.

1
The Rise of Fascism
A World in Turmoil

Prelude to the rise of fascism

In the aftermath of the First World War, Europe faced unprecedented challenges, leading to a period of profound socio-political upheaval. Economic turmoil, acute social discontent and geopolitical uncertainty provided fertile ground for the emergence of radical ideologies, laying the foundations for the rise of fascism across the continent. In Italy, the rise of Benito Mussolini's National Fascist Party paved the way for a totalitarian regime that sought to restore national pride amid economic distress. Simultaneously, in Germany, the seeds of discontent sown by the Treaty of Versailles germinated in an unstable environment, providing fertile ground for the rise of Adolf Hitler's National Socialist German Workers' Party. Here, we highlight the complex interplay of historical factors and underlying ideological currents that underpinned the rise of fascist movements. We examine the role of fervent nationalism, economic disillusionment, and manipulation of public sentiment as precursors to the erosion of democratic institutions and the consolidation of authoritarian rule. We examine the impact of charismatic leadership, political propaganda, and the exploitation of societal grievances to reinforce the appeal of fascist ideologies.

The evolving dynamics of international relations and the interconnectedness of global events are examined for their contribution to the proliferation of fascist doctrines beyond national borders. By exploring the socio-economic, political, and cultural landscapes that paved the way for the rise of fascism, we aim to offer a nuanced understanding of the historical trajectory that precipitated one of the most tumultuous periods of the 20th century. By unravelling the complexities underlying the genesis and spread of fascist ideologies, we seek to illuminate the intricate tapestry of factors that rooted these belief systems in the collective consciousness of societies, leading to profound implications for subsequent developments in world history.

The historical precursors of fascism

The roots of fascism can be traced back to a complex interplay of social, economic, and political factors that permeated the early 20th century. From a historical perspective, the aftermath of the First World War was marked by profound disillusionment, economic instability, and social unrest throughout Europe. The unprecedented scale of destruction and loss plunged many nations into despair, paving the way for the rise of radical ideologies. At the heart of this tumultuous period were the dire socio-economic conditions that afflicted post-war societies. This combination of circumstances created fertile ground for the emergence of fascist movements, which sought to exploit the prevailing discontent and uncertainty.

The inability of traditional political systems to address these challenges engendered a deep sense of disillusionment among the population, leading to an erosion of faith in established institutions. Economically, the aftermath of the First World War was marked by profound upheaval, with devastated economies struggling to recover amid rampant inflation and unemployment. The heavy reparations imposed on the defeated nations further exacerbated their plight, sowing the seeds of resentment and fostering an environment conducive to radical ideologies. These economic difficulties were accompanied by a widespread sense of political instability, marked by fractured governments, frequent changes in leadership and widespread disillusionment with democratic processes. Amid this turmoil, various groups espousing extremist ideologies capitalised on the prevailing discontent to gain support and spread their vision of radical transformation.

The social fabric of Europe was under considerable strain in the aftermath of the war, as societal upheavals and the trauma of conflict reshaped interpersonal dynamics and collective identities. The erosion of traditional social structures, combined with a climate of uncertainty and fear, fuelled widespread anxiety and insecurity among the population. In such a climate, charismatic leaders emerged to exploit these vulnerabilities, offering simplistic solutions and scapegoats for the complex challenges facing their societies. Consequently, the historical precursors of fascism can be discerned from this complex interplay of economic despair, political instability and social upheaval, highlighting the multifaceted nature of its emergence and spread.

Economic Despair and Political Instability

The interwar period was marked by a climate of economic despair and political instability throughout Europe, which set the stage for the rise of fascist ideologies. After the ravages of the First World War, nations found themselves grappling with paralysed economies, widespread unemployment, and hyperinflation. The reparations imposed on Germany by the Treaty of Versailles exacerbated its economic downturn, fuelling resentment and providing fertile ground for radical political movements. At the same time, Italy faced similar turmoil, characterised by economic recession, social unrest and disaffection with the ruling establishment. As political institutions struggled to address these challenges, social and cultural tensions increased, giving rise to ideological polarisation and heightened class conflict. The collapse of traditional power structures and the inability of democratic systems to provide stability created an atmosphere of uncertainty and disillusionment among the population. This climate of discontent paved the way for charismatic leaders who exploited the prevailing sense of vulnerability and offered radical solutions promising rapid economic recovery and national rejuvenation.

Economic despair was compounded by the simultaneous erosion of political stability. Governments faced internal divisions, legislative gridlock, and frequent changes in leadership, resulting in a loss of public confidence in established political processes. This environment of volatility and inefficiency gave rise to a desire for decisive leadership and strong governance, providing an opportunity for authoritarian figures to position themselves as saviours of the nation.

The global financial crisis of 1929, commonly known as the Great Depression, plunged the world into deeper economic turmoil and propelled many countries into social and political chaos. The collapse of financial institutions, mass unemployment and widespread poverty created a potent cocktail of fear, anger and uncertainty, exposing the vulnerabilities of existing socio-political frameworks. The resulting social upheaval provided fertile ground for the spread of radical ideologies, particularly those espousing nationalist fervour and promises of restored glory. In essence, the convergence of economic despair and political instability during the interwar period provided the backdrop for the emergence and rise of fascist movements in various European nations. Understanding this tumultuous landscape is essential to grasping the genesis and appeal of the fascist ideologies that swept across the continent during this pivotal period.

The Rise of Mussolini and Italian Fascism

In the chaos of post-World War I Europe, Italy faced economic challenges and political fragmentation. Exploiting the prevailing discontent, Benito Mussolini emerged as a formidable figure, laying the foundations for Italian fascism. Mussolini capitalised on the disillusionment stemming from Italy's perceived marginalisation in the Treaty of Versailles and the lack of substantial gains resulting from the war effort. The appeal of a revival of the glory days of the Roman Empire and strong leadership resonated with many disillusioned Italians, providing fertile ground for his rise to power. Mussolini's skilful use of propaganda and mass mobilisation through the Blackshirts, as well as promises of social order and rejuvenation of the nation, enabled him to gain significant support.

The creation of the National Fascist Party and the infamous March on Rome in 1922 marked a decisive turning point, culminating in his appointment as Prime Minister and the consolidation of his authority. The subsequent imposition of authoritarian measures served to suppress dissent and crush rival political forces. Under Mussolini's rule, Italy underwent a rapid transformation, marked by aggressive nationalism, expansionist ambitions and uncompromising repression of political opposition. The cult of personality surrounding Il Duce, combined with the glorification of militarism and traditional values, fuelled nationalist fervour among the population. This fervour was fuelled by orchestrated spectacles and propaganda demonstrations, reinforcing Italy's image as a resurgent world power.

The institutionalisation of fascist ideology led to aggressive corporatist policies and control over various sectors of society, including the economy, the media and education. The fascist regime cultivated a mixture of authoritarianism and totalitarianism, curtailing individual freedoms and subverting democratic institutions. At the same time, a brutal campaign against perceived internal opponents, exemplified by the Matteotti crisis, highlighted the ruthless nature of Mussolini's grip on power. In an unstable geopolitical context, Mussolini pursued his territorial ambitions, leading him to invade Ethiopia in 1935. This expansionist venture was intended not only to emulate the imperial conquests of ancient Rome, but also to strengthen Italy's position on the world stage. The ensuing international condemnation and the lukewarm response of the League of Nations underscored Italy's precarious position, foreshadowing the escalating tensions that would engulf the world. The rise of Italian fascism is a compelling testament to the seductive appeal of authoritarian regimes in times of uncertainty, laying bare the complex interplay of historical circumstances and individual action in shaping the destiny of nations.

Hitler's Rise in Germany

Adolf Hitler's rise to power in Germany is an important and complex chapter in the annals of history. In the turmoil of the First World War, a myriad of socio-economic factors paved the way for Hitler's rise. The Treaty of Versailles had plunged Germany into enormous economic hardship and political instability, providing fertile ground for a leader capable of exploiting feelings of discontent and disillusionment. Hitler's charismatic oratory skills and fervent nationalist rhetoric resonated with a population thirsty for change. His National-al Socialist German Workers' Party, or Nazi Party, quickly gained support by promising to restore Germany to its former glory. Through passionate speeches, strategic alliances, and exploitation of popular discontent, Hitler managed to rise to the chancellorship in 1933.

Once in power, Hitler quickly consolidated his authority, orchestrating a totalitarian regime that encompassed all aspects of German society. His unrivalled ability to manipulate propaganda and the media facilitated the spread of a twisted ideology, glorifying Aryan supremacy while demonising marginalised groups, particularly Jews, Roma and people with disabilities. The aggressive expansion of state control, exemplified by the creation of the Gestapo and the SS, illustrated the systematic erosion of civil liberties and democratic norms. Hitler's astute manipulation of foreign policy further bolstered his rise, exploiting geopolitical tensions to realise his territorial ambitions and consolidate his alliances. In retrospect, Hitler's rise serves as a cautionary tale, highlighting the catastrophic consequences that can arise when demagoguery and unchecked power intersect with societal dissonance.

Ideological Principles and Propaganda

The ideological principles and propaganda employed by the fascist regime were powerful tools for shaping public perception and consolidating power. At the core of fascist ideology was a belief in the supremacy of the state, an emphasis on nationalist fervour, and the glorification of a perceived superior race. Building on these principles, the regime sought to create a sense of unity and identity among the population, while marginalising minority groups and dissenting voices. Propaganda became an omnipresent force in spreading the fascist message, permeating various forms of media such as newspapers, radio broadcasts, films, and visual arts. The imagery and rhetoric used in propaganda served to deify the leader, demonise perceived enemies, and instil obedience to the regime's directives. By controlling the dissemination of information and manipulating public opinion, the regime aimed to foster unwavering loyalty and conformity among the population.

The propaganda apparatus played a crucial role in developing a cult of personality around the leaders, presenting them as infallible figures embodying the glory and aspirations of the nation. This mythologisation of the leaders helped consolidate their grip on power and reinforce the idea that they were indispensable to the nation's destiny. The meticulous orchestration of mass events, rallies and spectacles also served as powerful tools for spreading the regime's ideology and fostering a sense of belonging and usefulness among citizens. These grand demonstrations of power and unity were meticulously choreographed to instil fear, fervour and a sense of collective strength, effectively subsuming individual identity into the collective consciousness of the state.

The exploitation of modern technologies and mass communication allowed the regime to spread its ideals beyond geographical borders, projecting an image of invincibility and ideological righteousness on the world stage. The reach of fascist propaganda extended far and wide, aiming to captivate and influence international audiences, thus extending its influence beyond national borders. In short, the manipulation of ideology and the ubiquitous dissemination of propaganda were integral to the consolidation of power by fascist regimes. By shaping perceptions, instilling fervent loyalty, and encouraging the cult of personality, the regime sought to mould the collective consciousness of the population, establish its authority, and project its influence both nationally and internationally.

Consolidation of power: tactics and strategies

The consolidation of power by fascist regimes involved a complex network of tactics and strategies aimed at centralising authority and suppressing dissent. One of the main methods used by these regimes was to manipulate legal frameworks to justify repressive measures. Laws were enacted to restrict civil liberties, muzzle the press, and establish a pervasive surveillance system, creating an environment of fear and paranoia. Dissidents, intellectuals, and perceived enemies of the state were systematically targeted with arbitrary arrests, detention without fair trial, and punitive measures that effectively eliminated opposition.

Propaganda played a key role in consolidating power. State-controlled media and carefully crafted messages were used to instil unwavering loyalty to the regime, propagating an idealised image of the leader and demonising any dissenting voices. Propaganda was not limited to traditional media, but also permeated public spaces, schools, and cultural events, subverting independent thought and encouraging blind allegiance. Fascist regimes also relied on paramilitary forces and secret police to stifle resistance and maintain control. These organisations operated with impunity, instilling terror in the population and eradicating any semblance of opposition. The deployment of violence and intimidation tactics served as a brutal reminder of the consequences of challenging the established order, silencing dissent through coercion and fear.

Fascist leaders forged strategic alliances with influential industrialists, military leaders, and other representatives of power to consolidate their grip on key institutions and resources, thereby securing the loyalty of vital sectors of society. In addition, the creation of a cult of personality around the leader reinforced the consolidation of power. Their omnipresence in public imagery, political discourse, and daily life cultivated a sense of reverence and authority that transcended rational criticism. Ritualised displays of loyalty and adulation further reinforced the leader's hold on the collective psyche of the population, fostering an environment where unquestioning obedience was considered a virtue. It is essential to recognise that the consolidation of power by fascist regimes was not based solely on overt coercion and violence; rather, it was a calculated orchestration of legal, propaganda, paramilitary, and ideological mechanisms that effectively stifled opposing voices and cemented the authoritarian regime.

Impact on society and cultural institutions

The consolidation of power by fascist regimes inevitably led to profound transformations in societal structures and cultural institutions. As these authoritarian governments sought to control all aspects of public life, the impact on society and cultural institutions was both extensive and lasting. One of the most striking effects was the systematic suppression of intellectual freedom and artistic expression. Censorship became pervasive, and strict rules were imposed on literature, art, music, and academic discourse. Intellectuals, artists, and academics who did not conform to the ideological dictates of the state were often persecuted, exiled, or imprisoned. This repression stifled innovation and creativity, fracturing the cultural landscape and relegating once-vibrant centres of intellectual exchange and artistic experimentation to the shadows.

Fascist ideology sought to redefine society's norms and values, imposing a rigid framework that restricted individual freedoms and marginalised minority groups. The persecution of ethnic and religious minorities, along with the enforcement of gender roles, led to widespread fear, discrimination and disenfranchisement. These social upheavals irreversibly altered the fabric of communities and shattered the common sense of identity and belonging. At the same time, the state vigorously propagated its own narrative and mythologised national history and heritage, using propaganda mechanisms to indoctrinate the population and cultivate unwavering loyalty to the regime. Through mass rallies, symbolism and carefully crafted narratives, the ruling elite sought to forge a unified collective consciousness, aligning cultural production and historical narratives with its political agenda. The repercussions of these coercive mechanisms have reverberated across generations, leaving an indelible mark on the social psyche and cultural memory of the societies affected. The erosion of critical discourse, the subjugation of marginalised voices and the manipulation of cultural objects for ideological purposes are lasting legacies that inspire subsequent generations to address the deep scars left by this dark chapter in history.

International reactions and early resistance

International reactions to the rise of fascism were highly diverse, reflecting the complex geopolitical landscape of the time. While some nations were alarmed by the rapid rise of authoritarian regimes, others pursued policies of appeasement or non-intervention. The imposition of authoritarian control, propaganda and military expansion by the fascist powers provoked a variety of reactions from the international community. France and Britain, for example, initially sought to avoid confrontation with Germany and Italy through diplomatic negotiations and concessions, a strategy that ultimately proved unsuccessful as Hitler's ambitions became increasingly apparent. Meanwhile, in academic circles, the media and civil society, the first signs of resistance to authoritarian regimes began to emerge. Intellectuals, artists and activists from across Europe and beyond voiced their opposition to the spread of fascism and its impact on cultural and intellectual freedom. Organisations and individuals concerned with preserving democratic principles and human rights began to organise themselves into clandestine networks, laying the groundwork for future acts of resistance.

The growing flow of refugees fleeing persecution in fascist-controlled territories raised awareness of the plight of those oppressed under these regimes. The experiences and testimonies of these refugees galvanised support for anti-fascist movements and highlighted the urgent need for international solidarity. The accumulation of tensions and open defiance of fascist ideologies paved the way for the alliances and coalitions that would emerge in the context of a global conflict. International reactions and early resistance played a vital role in the unfolding of events that led to the outbreak of the Second World War, highlighting the power of collective action and moral courage in the face of the forces of tyranny and oppression.

Comparative analysis with other authoritarian regimes

As we delve into the complex layers of authoritarian regimes, it becomes imperative to conduct a comparative analysis in order to better understand their nature and impact. Fascism, although distinct in its manifestation, shares commonalities with other historical authoritarian systems spanning different cultural and geographical contexts. The rise of totalitarianism in the early 20th century is an essential subject for comparison, allowing parallels and distinctions to be drawn between fascist movements and their counterparts. Soviet communism under Joseph Stalin is a notable regime for comparative analysis. Although ideologically different from fascism, Stalinism bore striking similarities in terms of the cult of personality, centralisation of power, and repression of dissent. Propaganda mechanisms and state control of the media prevailed in both systems, fostering a totalitarian environment characterised by surveillance and ideological conformity.

Examining the rise of authoritarianism in Spain under Francisco Franco offers valuable insights into the pervasive influence of nationalist fervour and the militarisation of society – a hallmark of fascist governance. The autocratic structure of Franco's Spain, marked by the persecution of political opponents and rigorous censorship, bears parallels to the repressive tactics employed by Nazi Germany and Mussolini's Italy. In examining Latin American dictatorships, notably the reigns of Augusto Pinochet in Chile and Juan Perón in Argentina, we encounter authoritarian paradigms characterised by the imposition of a single-party state, violent repression of opposition, and prioritisation of nationalist agendas. These similarities prompt us to explore the transnational spread of dictatorial models that transcended continental boundaries, echoing the global reach of fascist ideologies.

Contemporary autocracies exhibit enduring characteristics reminiscent of historical fascist states, illustrating the perpetual resonance of authoritarian governance across temporal and geographical domains. The consolidation of power through indoctrination, mass mobilisation, and the cult of leadership underscores the enduring legacy of totalitarian ideologies. In sum, a comparative analysis with other authoritarian regimes helps elucidate the interconnections and divergences in the historical trajectory of oppressive systems. By placing fascism within the broader context of authoritarianism, we gain a holistic understanding of its emergence, consolidation of power, and lasting impact—a nuanced perspective essential for discerning the complexities of authoritarian rule and its implications in contemporary societies.

The transition to the prelude to war

As fascist regimes tightened their grip on Europe, the transition to the prelude to war became undeniable. The inherently expansionist nature of these authoritarian governments, coupled with their aggressive foreign policy, sowed the seeds of imminent conflict. The backdrop of economic depression and increased militarisation further fuelled growing tensions between nations. This era was marked by the erosion of diplomatic alliances, with fascist powers pursuing their path of conquest through military force and strategic coercion. The alliances and treaties that had preserved the fragile peace in the aftermath of the First World War began to unravel. The resurgence of nationalism, coupled with fervent ideologies of supremacy, created an atmosphere conducive to territorial ambitions and power struggles. This sense of entitlement to domination exacerbated geopolitical rivalries, sparking a relentless quest for territorial expansion.

Simultaneously, the rise of fascism sparked widespread apprehension and triggered international debate and intervention. Policymakers were confronted with the complex interplay of military strategies, economic imperatives, and moral obligations. The arms race and rapid rearmament initiatives further destabilised the global balance and set the stage for impending cataclysm. The tangle of secret alliances, clandestine negotiations, and covert operations highlighted the underhanded manoeuvring of influential world powers. As diplomatic channels ran dry, the momentum toward war gathered pace, plunging the international landscape into uncertainty and unease. The crescendo of militaristic fervour reverberates across the continent, while the palpable fear of conflict is palpable. The pervasive spectre of war has permeated every aspect of societal discourse and mobilised populations on an unprecedented scale.

Ultimately, the transition to the prelude to war stands as a harrowing testament to the fragility of global stability and the dangers of unchecked aggression. It symbolises a pivotal moment in history, where the forces of totalitarianism and democratic ideals clashed with grave consequences. This tumultuous period is a grim reminder of the disastrous repercussions of political extremism and unbridled expansionism, which shaped the trajectory of an era marked by turmoil and upheaval.

2
Operation Paperclip
The Recruitment of Genius

Operation Paperclip

Operation Paperclip is a poignant testament to the complex interplay between scientific progress, political strategy and ethical considerations in the aftermath of the Second World War. The primary objective of this secret US government initiative was to exploit the intellectual prowess of German scientists, particularly those who had been involved in cutting-edge technological and scientific research during the war. Underlying this ambitious undertaking was the strategic imperative to gain a competitive advantage in the escalating Cold War, where the use of cutting-edge expertise promised decisive advances in military capabilities, space exploration and industrial innovation.

Operation Paperclip was born out of a combination of pragmatic geopolitical calculations and an ongoing quest for knowledge acquisition. The contributions of German scientists during the war were revered and coveted, leading to a fervent quest to secure their expertise to advance American interests in the post-war period. Although the moral implications of employing former adversaries raised compelling ethical dilemmas, the perceived need to stay ahead of technological advances served as justification for the programme's implementation. The operational mechanics of Operation Paperclip relied on the meticulous selection of prominent scientific figures whose collective knowledge and skills were perceived as invaluable assets to the trajectory of American scientific endeavours. This selective process resulted in the recruitment of renowned minds in various disciplines, including physics, engineering, medicine, and aerospace research.

By examining the historical background that catalysed the formation of Operation Paperclip, we gain crucial insights into the intertwined narratives of scientific research, national security imperatives, and the ethical tightrope walk of decision-makers navigating the aftermath of a world war. As we delve deeper into the annals of Operation Paperclip, it becomes clear that the intersection of ambition, competition, and moral dilemmas gave rise to a groundbreaking chapter in the annals of scientific history. Against a backdrop of ideological tensions and the demands of an evolving world order, the integration of German scientific luminaries into the fabric of American research institutions sparked a confluence of progress and controversy, leaving an indelible mark on the tapestry of post-war scientific developments.

Historical Context and Origins

After the end of World War II, the race for scientific supremacy became a central feature of the global political landscape. In the aftermath of the war, world powers recognised that scientific innovation was the key to military dominance and economic prosperity. It was in this climate of intense competition and urgency that Operation Paperclip began to take shape. This secret initiative, born out of the United States' fervent desire to gain an advantage in the nascent Cold War, aimed to exploit the intellectual prowess of German scientists, particularly those working in the fields of advanced weaponry and aerospace technology.

The historical roots of Operation Paperclip date back to the end of the war, as Allied forces advanced into Nazi-occupied territories. The prospect of obtaining the expertise of Nazi scientists became increasingly appealing to American intelligence services. Fearing that the Soviet Union would acquire high-level scientists, the United States launched a clandestine operation to identify, recruit, and transfer these individuals to American soil, to protect them from prosecution for their role in the Nazi regime. This complex undertaking was underpinned by a set of geopolitical considerations, ethical dilemmas and pragmatic calculations. The harrowing experiences of the war, combined with global geopolitical ambitions, converged to form the backdrop against which Operation Paperclip came into being.

The project originated in a deep fear that an ideological adversary would gain technological ascendancy and strategic advantage. Such a context underscores the extent to which scientific capabilities were not only equated with military power, but also linked to ideological superiority and diplomatic influence. Operation Paperclip thus embodies the intersection of history, ideology, and technological ambition. Its roots run deep in the annals of post-war power dynamics, encapsulating the complex interplay between national interests, wartime exigencies, and the moral compromises made in the pursuit of scientific progress. This chapter of history is a poignant testament to the tangled legacy of human achievement and moral complexity, offering a compelling narrative of ambition, aspiration, and the lasting consequences of scientific endeavour in the crucible of conflict.

Identification and Selection of Scientists

Operation Paperclip involved the meticulous identification and selection of scientists and specialists from Nazi Germany. The process included a comprehensive assessment of individuals with scientific expertise, particularly in areas related to military and technological advancement.

A special team composed of intelligence officers, military officials, and researchers was entrusted with this critical responsibility, knowing that recruitment decisions would significantly influence the trajectory of post-war technological innovation.

The identification phase required an exhaustive analysis of scientific achievements, academic credentials, and professional networks in Germany. Working closely with informants and defectors, the task force sought to uncover hidden gems of scientific acumen amid the turmoil of war. The wartime landscape presented many challenges, as chaos and confusion often obscured the true potential of these individuals. However, through persistent investigation and strategic alliances, the task force gradually gained a clearer picture of the scientific prowess available for assimilation into the Allied scientific community. The selection criteria encompassed not only technical skills, but also ideological alignment and a willingness to contribute to the reconstruction of the global scientific domain. With an emphasis on sustainable progress and ethical considerations, the task force strove to ensure that the scientists selected embodied values compatible with the democratic principles espoused by the Allies. This delicate balance between scientific excellence and ethical integrity highlighted the challenging nature of the selection process, which sought to navigate the moral complexities inherent in harnessing the potential of former adversaries.

The selection of scientists went beyond individual merit to encompass potential for collaboration and interdisciplinary synergy. By emphasising a holistic approach to the acquisition and dissemination of knowledge, the working group recognised the importance of bringing together diverse skills that could collectively propel scientific frontiers into uncharted territory. These deliberations transcended national boundaries, reflecting a resolute commitment to prioritise scientific progress over geopolitical rivalries. Ultimately, the identification and selection of scientists under Operation Paperclip was a momentous undertaking that profoundly shaped the course of post-war scientific collaboration. The fusion of intellect, ethics, and ambition paved the way for unprecedented breakthroughs, heralding a new era of innovation forged by the convergence of diverse scientific legacies.

Legal framework and secret agreements

After the identification and selection of scientists under Operation Paperclip, it was imperative for the US government to establish a legal framework and conclude secret agreements to facilitate the transfer of expertise and knowledge. The complexity of this process was deeply rooted in Cold War politics and the race for technological superiority. The legal framework mainly encompassed immigration and naturalisation laws, security clearances and contractual obligations. By establishing specific protocols, the US government sought to reconcile the recruitment of German scientists with its immigration policies, ensuring that these specialists could contribute to American scientific and technological advances. The delicate balance between national security and scientific progress required rigorous screening processes and thorough background checks.

Legal obstacles to the entry of foreign nationals into the country, particularly those who had been aligned with Nazi Germany, had to be overcome. Secret agreements were made to keep Operation Paperclip confidential and out of the public eye or the international community. These agreements required the utmost discretion and cooperation between the American government and the recruited scientists, binding them to strict confidentiality regarding their participation in the operation. The clandestine nature of these agreements underscored the sensitivity of the project and the importance of securing the intellectual capital acquired. These agreements helped preserve the reputation of the scientists involved, protecting them from potential negative reactions or persecution. In addition, they aimed to mitigate potential diplomatic repercussions while preserving the integrity of the collaborative efforts between the United States and Germany. The complex legal and contractual framework of Operation Paperclip highlights the ethical, moral, and legal dilemmas faced by this secret initiative. It reflects the unprecedented intersection of geopolitics, law and science, highlighting the transformative impact of historical events on the development of contemporary scientific endeavours.

Key figures and pioneers

The success of Operation Paperclip relied heavily on the individuals involved, whose expertise proved invaluable to the development of various fields after the war. Among the major figures and pioneers brought to the United States as part of this programme were prominent scientists, engineers and researchers such as Wernher von Braun, a talented rocket scientist who played a pivotal role in advancing American space exploration through his work with NASA. Von Braun's leadership and technical insight were instrumental in the development of the Saturn V launch vehicle, which propelled the Apollo missions to the moon and cemented the United States' dominance in the field of space technology. Arthur Rudolph is another pioneer, renowned for his contribution to the development of the Saturn V rocket and the American ballistic missile programme, which greatly enhanced the nation's strategic defence capabilities. These individuals, along with many others, played a vital role in reshaping the scientific and technological landscape of the United States, leaving an indelible mark on history and progress. Their expertise not only contributed to national security and aerospace projects, but also laid the foundation for revolutionary advances that continue to benefit society as a whole. The legacy of these major figures and pioneers highlights the complex interplay between scientific prowess, geopolitical strategies, and ethical considerations, prompting profound reflections on the lasting impact of their contributions.

Technology Transfer: From Warfare to Well-being

At the end of the Second World War, the United States found itself in possession of a treasure trove of technological advances and scientific expertise thanks to Operation Paperclip. This influx of knowledge and innovation led to a paradigm shift, as these wartime technologies were repurposed for more peaceful and constructive ends. The transformation of war-focused technologies into tools for the well-being of society represented a turning point in human history. The transfer of these advances enabled rapid progress in fields ranging from medicine to space exploration. Former military technologies were repurposed to improve civilian life, enabling breakthroughs in areas such as public health, infrastructure development, and environmental preservation. A notable example of this transfer is the use of rocket technology. Originally developed for military applications, rockets have become the cornerstone of space exploration, supporting scientific research and international cooperation. This shift highlights the immense potential of redirecting wartime innovations towards peaceful projects, demonstrating the adaptability and resilience of human ingenuity.

The transition from war to wellbeing has also reshaped strategic alliances and global geopolitics, encouraging collaboration and diplomacy instead of conflict and confrontation. The shared pursuit of knowledge and progress has facilitated mutual understanding between nations and cultures, paving the way for a more harmonious and interconnected world. However, the process of redirecting war technologies towards peaceful activities has not been without its problems and ethical questions. The dual-use nature of many of these innovations has raised questions about their purpose and potential implications. Striking a balance between harnessing the capabilities of these advances for the betterment of society and preventing their use for destructive purposes has posed a profound dilemma, requiring careful oversight and ethical discernment. As we delve deeper into the narrative of war and wellbeing, it becomes clear that technology transfer transcends mere scientific progress; it is a testament to humanity's adaptability and ingenuity. This evolution is a poignant reminder of the transformative power of innovation and the responsibility that comes with its deployment.

Challenges and controversies

When delving into the depths of Operation Paperclip, one cannot ignore the myriad challenges and controversies that surrounded this secret initiative. As the United States sought to harness the scientific expertise of German and Austrian researchers for post-war progress, it faced numerous ethical dilemmas and operational obstacles. One of the main challenges was the scrutiny and discontent expressed by some members of the scientific community and the public, who viewed the recruitment of former scientists affiliated with the Nazis as a betrayal of moral principles. This situation sparked heated controversy, with debates raging over whether the potential benefits of exploiting German intelligence outweighed the moral repugnance of employing individuals with ties to the Nazi regime.

The validity of the intelligence reports used to assess the value of each scientist was questioned, raising questions about the credibility and transparency of the selection process. The use of former Nazi technology for peacetime innovation also sparked heated discussions about the true intentions behind the acquisition of these assets and the potential implications for global power dynamics.

The integration of foreign scientists into American society and research institutions posed cultural and bureaucratic challenges, with language barriers, the ability to adapt to new norms, and different scientific methodologies posing obstacles to seamless collaboration. Furthermore, the absolute secrecy surrounding Operation Paperclip led to internal tensions within government agencies and conflicts of interest, fuelling suspicion and dissent within the ranks. These contentious elements cast a shadow of doubt and unease over the project, raising complex ethical, political, and practical dilemmas that required careful navigation. The controversies surrounding Operation Paperclip continue to prompt deep reflection on the intersection of science, ethics, and national interests, reminding us of the complex and challenging nature of scientific diplomacy in a context of geopolitical upheaval.

Cultural and Ethical Implications

Operation Paperclip, despite its undeniable contributions to technological progress, has been the subject of intense controversy due to its cultural and ethical implications. The recruitment and integration of German scientists into the American research and development sector raised significant moral and political dilemmas that continue to reverberate throughout history.

At the heart of these debates is the ethically ambiguous decision to pardon and employ individuals directly associated with the atrocities of the Nazi regime. The cultural clash between wartime adversaries, combined with the scientific expertise they possessed, created a complex and contentious landscape.

The ethical considerations surrounding the use of technologies developed by individuals complicit in war crimes prompt deep reflection on the responsibility of nations and individuals in exploiting innovations with such morally tainted origins. The echoes of Operation Paperclip resonate as a warning about the eternal conflict between progress and principles. This historical episode prompts us to reflect on the delicate balance between scientific progress and ethical integrity, reminding us that every step towards progress requires careful consideration of its potential consequences. Thus, the cultural and ethical implications of Operation Paperclip extend beyond its immediate impact and remind us that humanity's progress comes with lasting ethical responsibilities.

Post-war scientific contributions

After the tumultuous years of the Second World War, Operation Paperclip facilitated the transfer of remarkable scientific minds from war-torn Europe to the United States, stimulating research and innovation in various fields. These brilliant minds played a vital role in reshaping and advancing key scientific disciplines, leaving an indelible mark on the future of technology, medicine, and space exploration. In the field of rocketry and aviation, the influx of German scientists and engineers significantly accelerated the development of aerospace technology. Their expertise and ideas paved the way for monumental achievements such as the Apollo moon landing and the exploration of outer space. The legacy of these post-war scientific contributions continues to reverberate in modern advances in propulsion systems, satellite communication, and planetary exploration.

The medical community experienced profound advances thanks to the influx of exiled scientists. Breakthroughs in pharmacology, infectious disease control, and surgical techniques were fuelled by the wealth of knowledge and expertise brought by these esteemed individuals. Their contributions not only revolutionised medical practices, but also laid the groundwork for current research in areas such as immunotherapy, genetic engineering, and personalised medicine. In the field of theoretical physics and nuclear science, the contributions of exiled scientists proved transformative. Their collective efforts led to breakthroughs in nuclear energy, particle physics, and quantum mechanics, fundamentally altering our understanding of the universe. The implications of their work extended beyond scientific research, shaping global energy policies and encouraging international cooperation in nuclear research and non-proliferation initiatives.

The interdisciplinary collaborations sparked by Operation Paperclip catalysed innovations in various sectors, from materials science to computer science. The influx of scientific talent fostered a culture of intellectual exchange and cross-fertilisation, leading to paradigm-shifting developments in fields such as information technology, nanotechnology, and sustainable engineering. The influence of these post-war scientific contributions transcends temporal and geographical boundaries, permeating contemporary technological landscapes and societal frameworks. By reflecting on the lasting impact of these visionary individuals, we gain a deeper understanding of the complex interplay between historical events and scientific progress, highlighting Operation Paperclip's enduring legacy on the trajectory of human knowledge and innovation.

Conclusion: Reflection on Impact

Operation Paperclip is a pivotal moment in history that shaped the trajectory of scientific progress and geopolitical strategies. Reflecting on the impact of this clandestine endeavour, it becomes evident that the integration of German scientists into the fabric of Western science and industry had far-reaching repercussions in many fields.

The post-war scientific contributions catalysed by Operation Paperclip not only enriched the fields of technology, medicine and defence, but also influenced the socio-political landscape of the time. When examining the ethical and moral dilemmas arising from Project Paperclip, it is essential to recognise the complex web of circumstances surrounding the global dynamics of the post-war period. The imperative for technological progress in the context of emerging Cold War tensions precipitated the absorption of expertise from former adversaries. This ethical conundrum continues to spark debate and reflection on the trade-offs between progress and moral rectitude.

The lasting impact of the recruited scientists' work underscores the interconnectedness of global innovation. Their discoveries and advances transcended national boundaries, fuelling scientific revolutions that reverberated around the world. From breakthroughs in rocket and aeronautics to medical research and engineering, the legacy of Project Paperclip remains a testament to human ingenuity and collaboration. The historical legacy of Operation Paperclip prompts contemporary society to contemplate the enduring lessons it offers.

By reconciling the shared histories of scientific enterprise and geopolitical manoeuvring, we are compelled to confront the orchestration of power dynamics and the nuanced interaction between knowledge and authority. The ethical nuances and historical precedents established by Project Paperclip serve as an indelible reminder of the intersecting forces that shape the course of human progress. Ultimately, reflecting on the impact of Operation Paperclip compels us to question the complex interrelationships between ambition, ethics, and consequences in the realm of scientific discovery. As we navigate the ever-changing landscape of technological innovation, the echoes of Project Paperclip resonate both as a cautionary tale and as a testament to the perpetual quest for knowledge and progress.

3
The Shadowy Collaboration

Origins and early conspiracies

The origins of clandestine collaboration between various entities during the tumultuous period of the Second World War are deeply rooted in the complex geopolitical landscape of the time. As the war raged and world powers vied for supremacy, clandestine agreements were made and secret alliances were formed to gain strategic advantages. These collaborations were often the result of a powerful mix of necessity, mutual interests, and ideological alignments that transcended national borders. The early conspiracies and networks that emerged during this period laid the groundwork for a significant impact on technology, intelligence, and global politics. Motivated by the desire to gain an advantage in the war effort, nations and factions engaged in clandestine negotiations and shared their technological expertise under a veil of secrecy.

The convergence of scientific advances, military strategy, and espionage created an environment where collaboration flourished in the shadows. The initial formation of these alliances was marked by a web of intrigue, with key players manoeuvring to secure their position and expand their sphere of influence. While some collaborations arose from pragmatic considerations, others were guided by ideological affinities and shared worldviews. The intertwining of geopolitical calculations and clandestine agendas paved the way for a chapter in history defined by secret partnerships and discreet exchanges of knowledge and resources. The early conspiracies, often shrouded in mystery and subterfuge, sowed the seeds of a complex network that would cast a long shadow over the post-war era, significantly influencing the trajectory of global politics and technological advances. Understanding the origins and motivations behind these clandestine collaborations reveals a captivating narrative that illustrates the intersection of ambition, power, and secrecy in the context of a world engulfed in conflict.

Definition of collaboration agreements

Collaboration agreements made during periods of conflict and tension between nations have played a vital role in the course of history. These agreements represent a complex network of alliances, understandings, and discreet negotiations that often remained hidden from public view for decades. At the heart of these agreements were profound strategic objectives, ranging from gaining technological advantages to acquiring intelligence knowledge. The specific terms of these collaborative pacts varied considerably, reflecting the complex interplay of political, military, and industrial interests.

These agreements transcended national boundaries and brought together unlikely partners driven by the common goal of gaining an advantage in war and beyond. Collaborative agreements were characterised by their clandestine nature and required a high level of confidentiality and discretion. These pacts involved meticulously crafted legal frameworks and operational protocols, outlining the scope of information exchanges, resource sharing, and joint development initiatives. These agreements formed the backbone of large-scale operations designed to leverage the expertise, resources, and technologies of multiple parties to achieve mutual objectives.

Not only did these agreements facilitate the transfer of essential knowledge, but they also paved the way for long-term partnerships that extended well beyond the conflicts in which they originated. Key elements of these collaborative agreements included delineating responsibilities, establishing channels of communication, and identifying areas of mutually beneficial cooperation. Each party brought unique capabilities and assets to the table, resulting in synergistic agreements that enabled rapid advances and breakthroughs in various fields.

The agreements reflect the convergence of distinct ideologies and national interests, highlighting the complex balance of power and pragmatism that underpin collaborations. It is essential to recognise that these collaborative agreements were not without challenges and complexities. They often gave rise to complex moral and ethical dilemmas, particularly with regard to the acquisition and use of dual-use knowledge. Furthermore, the dynamics of these agreements were influenced by ever-changing geopolitical landscapes and shifting alliances, adding layers of unpredictability and volatility to collaborative efforts. As we navigate the annals of history, elucidating the nuances of these agreements provides insight into the interconnectedness of world affairs, the dynamics of innovation, and the lasting repercussions of wartime collaboration.

Key Figures and Secret Dialogues

Collaboration between key figures from various organisations during the tumultuous period of post-war socio-political upheaval was shrouded in secrecy and clandestine dialogue. The convergence of minds and ideologies in the shadow of shared interests and objectives gave rise to a network of secret discussions and strategic liaisons that significantly shaped the course of history. Among these enigmatic figures were prominent scientists, intelligence agents, influential politicians, and industry magnates, whose participation in secret dialogues had a lasting impact on the global landscape. These individuals acted with discretion, often using coded communications and secure meeting places to protect their exchanges from scrutiny. The story of this secret collaboration reveals a network of complex conversations held in secret locations, during which crucial decisions were made behind closed doors. These dialogues were marked by astute negotiations, the exchange of exclusive knowledge, and the orchestration of influential pacts aimed at preserving mutual benefits and consolidating power.

The influence of these clandestine conversations extended beyond traditional diplomatic circles, permeating the realms of cutting-edge technology, covert operations, and strategic intelligence gathering. The depth of these interactions demonstrates the determination of key figures to navigate the complex tapestry of post-war geopolitics and technological advancement, often operating on the fringes of conventional dialogue. As the chapters of history unfold, the true extent of the impact of these secret dialogues continues to be a subject of intrigue and intense scholarly examination, shedding light on the hidden actions of key figures who wielded significant influence behind the veiled curtains of secrecy. The intertwined narratives of these enigmatic conversations not only provide insight into the complexities of post-war collaboration, but also serve as a poignant reminder of the profound implications of such interactions on a global scale.

Strategic exchanges: Technology and intelligence

Cooperation between these behind-the-scenes collaborators extended beyond the simple exchange of information. Strategic exchanges of technology and intelligence formed the backbone of their alliance, altering the course of history. This secret partnership facilitated the transfer of advanced scientific knowledge, sophisticated weapon designs, and cutting-edge technological innovations. The clandestine nature of these exchanges enabled the seamless acquisition of crucial intelligence, allowing each party to gain a competitive advantage in an environment marked by espionage and counter-espionage. Leaders and experts from both sides engaged in confidential negotiations, fuelling rapid advances in military capabilities and scientific breakthroughs. These interactions paved the way for unprecedented advances in aviation, rocketry, cryptography and industrial production, ultimately influencing the outcome of wartime conflicts and post-war developments. It is worth highlighting the orchestrated collaborations that blurred national boundaries, transcended political ideologies, and emphasised the shared ambition of unrivalled technological supremacy. Strategic exchanges were marked by a calculated blend of resources, expertise, and operational methodologies, with enigmatic collaboration seeking not only to surpass adversaries but also to shape the future trajectory of warfare and technological dominance. As the turmoil of global conflict unfolded, the interconnected network of technology and intelligence fostered an environment where innovation thrived in secrecy and covert manoeuvres. However, the consequences of these exchanges reverberated far beyond the era of war, leaving an indelible mark on the subsequent dynamics of industry, innovation, and international relations.

Covert Operations: Facilitation and Execution

Covert operations played a vital role in facilitating and executing clandestine collaboration between various entities during a tumultuous period in history. These clandestine activities involved the strategic acquisition of advanced technologies, the gathering of intelligence, and the orchestration of complex espionage missions that shaped the trajectory of technological progress. Operating behind a veil of secrecy, these covert operations were meticulously planned and executed with precision, often blurring the lines between ethical boundaries and the pursuit of strategic advantage. Facilitating covert operations involved establishing complex networks, operational logistics, and recruiting highly skilled agents with specialised skills in infiltration, surveillance, and sabotage. Executing these covert operations required a meticulous blend of strategic foresight, operational cunning, and deep knowledge of the geopolitical landscape. From covert supply missions to stealth reconnaissance efforts, these operations encompassed a wide range of clandestine activities conducted in the shadows of global conflict. The success of these operations depended on maintaining absolute discretion, exploiting the weaknesses of rival organisations, and bypassing multiple levels of security protocols.

The adaptability and resilience of the agents deployed on these covert missions were paramount and often determined the success or failure of the overall objectives. Furthermore, the convergence of political manoeuvring and technological subterfuge heightened the stakes of these covert operations, marking a crucial chapter in the annals of history. The seamless integration of cutting-edge technologies into the fabric of espionage highlighted the multifaceted nature of these operations, which transcend conventional warfare and redefine the dynamics of power games at the highest echelons of influence. While these covert enterprises unfolded in the shadows, they left a lasting impact on the course of history, shaping the narrative of international relations and technological innovation for decades to come. Unveiling the intricacies of these covert operations reveals a compelling saga of ingenuity, risk, and allegiance, illuminating the enigmatic undercurrents that influenced the trajectory of technological progress and global domination. The legacy of these covert operations continues to resonate in contemporary discourse, prompting profound reflections on the ethical conundrums and strategic imperatives entangled in the clandestine tapestry of collaborative endeavours.

The Role of Espionage in Technological Progress

Espionage, a clandestine art refined throughout history, played a vital role in technological progress during and after World War II. By unveiling the veiled stratagems and secret manoeuvres of various intelligence agencies, we will explore the complex interaction between espionage and technological progress. Under the guise of covert and backstabbing operations, intelligence units engaged in elaborate plans to infiltrate enemy territories and appropriate cutting-edge plans, expertise, and innovations. Through this subterfuge, they gained invaluable knowledge of their adversaries' scientific and industrial developments, thereby influencing the trajectory of the global technological landscape.

The wealth of stolen knowledge and surreptitiously gathered data led to breakthroughs in fields ranging from aerospace and weaponry to cryptography and medicine. The invaluable information gleaned through espionage not only stimulated the technological prowess of the Allies, but also reshaped the paradigm of post-war research and development. This fusion of clandestine activities and technological pursuits contributed to an era of unprecedented scientific innovation, shaping the course of modern civilisation. What transpired in the shadows of espionage continues to reverberate in contemporary technological advances, demonstrating the inextricable intertwining of espionage and technological progress.

Impact on the dynamics of war and the reactions of the Allies

The impact of clandestine collaboration on the dynamics of war and the reactions of the Allies was profound and multifaceted. Technological advances and intelligence gleaned from clandestine operations significantly altered the course of the war. One of the most notable influences was the integration of advanced German technology into Allied military strategies, which led to a fundamental shift in the dynamics of global conflict. This infusion of expertise and innovation brought about a paradigm shift in the nature of warfare, paving the way for the modernisation of military tactics and weaponry.

The Allies' responses to the revelations and implications of this collaboration were numerous and complex. Initially, there was an urgent need to exploit the newly acquired knowledge and capabilities while mitigating the potential threats posed by the enemy's advances. This led to a wave of strategic assessments and re-evaluations of defence mechanisms, which ultimately reshaped the landscape of global security and diplomacy.

The collaborative exchange of intelligence and technology triggered a transformative ripple effect across various sectors of national defence and geopolitical strategies. This has spawned unprecedented cooperative efforts among allied nations, fostering new alliances and operational frameworks aimed at exploiting and leveraging shared intelligence and innovations. These initiatives laid the groundwork for modern concepts of collective security and mutual defence that continue to shape international relations today.

The impact extended beyond the immediate military sphere, permeating socio-political landscapes and economic domains. The knowledge and advances gained catalysed industrial breakthroughs and facilitated the emergence of new industries, creating a lasting legacy of technological progress and economic growth. They also reinforced the discourse on the ethical and moral dimensions of scientific and technological advances, prompting introspection and international dialogue on regulations and oversight. In short, covert collaboration had a significant influence on the dynamics of war and prompted resolute responses from the Allies, leading to a redefinition of the world order. Its lasting impact was evident in revolutionary advances in the military, technological, and geopolitical spheres, fundamentally shaping the course of history and forging a legacy that continues to resonate in contemporary strategic considerations.

Controversies and ethical conflicts

The collaboration between former enemies during and after the Second World War to leverage scientific knowledge and technological expertise sparked significant controversy and raised complex ethical conflicts. While the alliance enabled unprecedented advances in various fields, it also plunged the nations involved into debates about the moral implications of their actions. One of the most controversial issues concerned the use of scientific results from experiments conducted by the Axis powers under questionable ethical standards. The Allies' decision to clandestinely recruit scientists involved in such practices and to take advantage of their research raised many ethical dilemmas. This raised pertinent questions about the responsibility of these influential nations to uphold ethical standards, even in times of conflict and emergency.

The controversial use of the results of inhumane experiments drew international criticism and condemnation. The strategic exchange of intelligence information within the collaborative network led to further ethical complexities. The deliberate concealment of vital discoveries and knowledge from the public for security reasons created a conflict between national security interests and transparency. This has impeded the free flow of information that could have contributed to global scientific progress and potentially prevented duplication of research efforts.

The lack of accountability and transparency surrounding these clandestine operations has led to deep-seated mistrust between countries and persistent scepticism about ulterior motives. Conflicting moral obligations and the pursuit of geopolitical advantages have led to a web of ethical dilemmas that continue to influence contemporary discussions on scientific integrity, international relations, and the role of governments in promoting responsible innovation. These ongoing controversies compel us to critically re-evaluate historical decisions and their implications, shaping our understanding of the complex intersection between science, morality, and power.

Declassification and public knowledge

The process of declassifying information that was once confidential or secret has long been a subject of controversy within governments and organisations around the world. In the wake of significant historical events, particularly those involving national security and defence operations, the release of classified documents is often subject to intense scrutiny and debate. That is why we are examining the complexities surrounding declassification and its impact on public knowledge. Through the declassification of certain documents, individuals and researchers gain access to a wealth of previously undisclosed information that serves as crucial resources for historical research and understanding. This access enables in-depth analysis and promotes a greater understanding of past events, including collaborative efforts and clandestine agreements that have shaped the trajectory of technological development and global conflicts.

The declassification process allows the public to scrutinise the actions and decisions of authorities, thereby contributing to transparency and accountability within government structures. While the release of this information sheds light on crucial historical events, it also raises several concerns and challenges. Declassification requires careful consideration of potential repercussions, particularly when sensitive documents may impact diplomatic relations, national security, or ongoing intelligence operations.

Selective declassification of documents may inadvertently perpetuate biases or incomplete narratives, thereby influencing the interpretation of historical events. Striking a balance between transparency and safeguarding essential interests remains an ongoing challenge in the declassification process. In the context of the Second World War and its aftermath, the declassification of documents relating to collaborative initiatives provides a better understanding of the complex network of international relations, scientific exchanges and the practical consequences of clandestine procedures. It offers critical insights into how public awareness and academic discourse contribute to the preservation and dissemination of knowledge about these pivotal periods in history.

As revelations continue to emerge through declassification efforts, societies strive to reconcile the implications of this new awareness with the need for sensitivity in handling classified information. The importance of declassification extends beyond the presentation of historical facts; it shapes contemporary discourse and encourages a nuanced assessment of ethically complex decisions. The prospect of expanded access to declassified documents underscores the imperative of navigating the intersection of historical clarity and responsible stewardship of sensitive data, affirming the ongoing duty to honour the integrity and enduring relevance of historically significant collaborations.

Long-term consequences and effects

Following the declassification of previously concealed information, the revelation of covert collaboration between wartime enemies had far-reaching consequences and long-term effects that reverberated across the geopolitical landscape. The public dissemination of these clandestine agreements sparked intense debate and scrutiny, reshaping historical narratives and post-war perceptions. The revelation of collaborative efforts challenged established beliefs about alliances and the moral implications of strategic partnerships, prompting a reassessment of wartime conduct and its impact on world affairs.

4
Rockets And The Space Race

Foundations and early visions

The early foundations of rocket science can be traced back to the profound visions and pioneering work of individuals who dared to explore the field of space and propulsion. In the early 20th century, Konstantin Tsiolkovsky, a Russian scientist, laid the theoretical groundwork for space exploration with his visionary concept of using rockets to travel beyond the Earth's atmosphere. Tsiolkovsky's revolutionary calculations and writings on rockets not only inspired future generations of scientists and engineers, but also became the foundation of modern astronautics. Alongside Tsiolkovsky's work, Robert Goddard, an American physicist, meticulously studied the principles of rocket propulsion and was the first to successfully launch a liquid-fuelled rocket in 1926. Goddard's innovative experiments and tireless dedication to perfecting rocket technology played a vital role in shaping the trajectory of space exploration.

These early pioneers faced scepticism and often encountered resistance in their quest for space travel. Their unwavering commitment to advancing rocket science, despite the many challenges they faced, laid the foundation for the monumental advances that would follow. The visions of Tsiolkovsky and Goddard were not limited to their respective countries, but sparked global interest in the untapped potential of space travel. Their work paved the way for a new era of scientific research and technological innovation, igniting a passion for conquering the final frontier. As we move forward in this historical narrative, it becomes clear that without the fundamental contributions of pioneers such as Tsiolkovsky and Goddard, the staggering advances made in rocket science and space exploration over the decades would never have been possible. These visionaries sowed the seeds of possibility at a time when the dream of reaching for the stars was seen as mere fantasy. Their enduring legacy continues to inspire and guide modern space exploration, testifying to the indomitable spirit of human ambition and ingenuity.

The Pioneers of Rocket Science

The early decades of the 20th century saw a boom in scientific exploration, with individuals becoming passionate about rocket science. At the forefront of this movement were several figures whose pioneering work laid the groundwork for the revolutionary advances that would follow. One such figure was Konstantin Tsiolkovsky, a Russian scientist and visionary whose theoretical work on rocket propulsion and space travel greatly influenced the development of astronautics. His revolutionary concept of using multi-stage rockets not only propelled the idea of interplanetary travel, but also inspired subsequent generations of scientists and engineers.

Meanwhile, in Germany, Hermann Oberth, often considered the father of the modern rocket, helped to elucidate the technical aspects of rocket design and propulsion. Through his writings and experiments, Oberth demonstrated the potential of liquid-fuelled rockets, paving the way for practical applications of space technology. Another key figure in rocket science is Robert H. Goddard, an American physicist and engineer whose extensive research and innovative experiments led to the first successful flight of a liquid-fuelled rocket in 1926. Goddard's decisive contributions to rocket design and his advocacy for space exploration earned him the title of 'father of the modern rocket'. Meanwhile, in France, the efforts of French physicist and engineer Jean-Jacques Barré stimulated European research in rocketry and astronautics. Barré's meticulous experiments with solid-fuel rockets demonstrated the potential of these propulsion systems, influencing subsequent developments in missile technology and space exploration. These visionaries, along with many other unsung contributors, collectively propelled rocket science into uncharted territory, sparking a global quest for knowledge and innovation that continues to shape our understanding of the cosmos.

World War II: Progress Under Pressure

During World War II, the race to develop advanced weapons and technologies reached an unprecedented level of urgency. As the conflict intensified, nations around the world devoted considerable resources to scientific research and engineering in an effort to gain a decisive advantage. This intense pressure led to remarkable advances in various fields, including rocket science. The need for long-range weapons capable of inflicting considerable damage spurred intense innovation in rocket technology. The Axis powers and the Allied powers sought to exploit the potential of rockets as weapons of war, leading to significant breakthroughs in the fields of propulsion and aerodynamics. In Germany, the Nazis pioneered the development of ballistic missiles, notably with the V-2 rocket, which became the world's first long-range guided ballistic missile. This technological feat revolutionised the concept of warfare and marked a turning point in the history of rocketry. The increased demand for rocket technology during the war enabled rapid advances that laid the foundation for subsequent space exploration. It is essential to recognise the complex ethical and moral implications of these developments, illustrated by the complicity of scientists in the militarisation of scientific progress.

The pursuit of military superiority inadvertently catalysed unprecedented scientific and technical achievements, laying the groundwork for the post-war space race and the exploration of the cosmos. The legacy of Second World War advances in rocket technology reverberates in modern innovations, underscoring the immense impact of the demands of war on the trajectory of scientific progress.

From V-2 rockets to space exploration

After the end of the Second World War, technological advances made in Germany, particularly through the development and deployment of the V-2 rocket, laid the foundations for a new era in human history: space exploration. The V-2 rocket, under the leadership of Wernher von Braun and his team, represented a leap forward in rocket technology, serving as a springboard for subsequent achievements in space travel. At the end of the war, the United States and the Soviet Union sought to capitalise on the knowledge and expertise gained from captured German scientists and the V-2 programme. This pivotal moment marked the beginning of intense competition between the two superpowers, sparking the race to conquer the final frontier.

Through Operation Paperclip, many German scientists, including von Braun, were brought to the United States to contribute their expertise to the burgeoning American rocket programme. At the same time, the Soviet Union also made considerable progress in rocket development, thanks to the contributions of its own captured German experts. The Cold War served as the backdrop for this intense rivalry, and the world witnessed fierce competition between the United States and the USSR, each seeking to outdo the other in the pursuit of space exploration. This period of transformation gave rise to remarkable achievements, culminating in landmark events such as the Soviet Union's successful launch of Sputnik 1, marking the dawn of the space age. These monumental developments not only propelled scientific progress, but also redefined the geopolitical landscape, shaping the future of international relations. From V-2 rockets to decisive breakthroughs in space exploration, this era laid the foundation for humanity's extraordinary journey beyond Earth, inspiring generations to reach for the stars.

The United States and the USSR became formidable rivals in the field of space exploration, sparking one of the most influential and intense competitions in human history. This unprecedented race highlighted the technological excellence and strategic prowess of both nations, shaping the course of scientific progress and geopolitical dynamics on a global scale. As the Cold War intensified, the launch of the Soviet satellite Sputnik in 1957 resonated around the world, marking the beginning of a fierce space rivalry. Spurred on by this Soviet achievement, the United States galvanised its efforts, leading to the creation of NASA in 1958. Subsequently, Yuri

Gagarin's historic flight as the first man in space in 1961 propelled the USSR into the limelight, heightening the sense of urgency surrounding the American space programme.

This competitive atmosphere gave rise to a series of monumental events, including President John F. Kennedy's bold commitment to land a man on the moon before the end of the 1960s. This statement embodied not only the fervent determination of the United States, but also the indomitable spirit of human exploration. The Apollo 11 adventure, which culminated in Neil Armstrong's iconic step onto the lunar surface in 1969, stands as a lasting testament to the unwavering determination and ingenuity of the American space enterprise. Furthermore, the significant achievements of the USSR cannot be overlooked, with milestones such as the Luna programme, which led to the success of the first unmanned robotic missions to the moon. This pioneering feat cemented the Soviet Union's position as a leading force in space science and engineering, inspiring further breakthroughs that advanced our understanding of celestial bodies. The competitive nature of the space race not only amplified the pace of technological innovation, but also generated international anticipation and captured the imagination of millions. The juxtaposition of the United States and the USSR symbolised the contrast between two distinct ideologies, shaping public opinion and fostering a collective sense of pride and wonder. Ultimately, this period of fervent competition laid the groundwork for transcendent advances in human achievement, stimulating limitless possibilities and fuelling progress that continues to resonate in our exploration of outer space. The legacy of the space race between the United States and the USSR remains a lasting testament to the indelible impact of determination, collaboration, and the unwavering pursuit of knowledge.

Technological innovations and their proliferation

The space race between the United States and the USSR ushered in an era of monumental technological advances. Technological innovations proliferated rapidly as both nations sought to outdo each other in reaching new milestones in space exploration. This intense rivalry sparked a wave of investment in research and development, fostering unprecedented advances in aerospace engineering and related fields. One of the most important technological innovations of this period was the miniaturisation and optimisation of electronics for space applications. Both superpowers devoted resources to developing compact yet durable electronic components capable of withstanding the rigours of space travel. This led to revolutionary achievements such as the development of guidance systems, communications satellites and sophisticated instruments essential for controlling spacecraft and transmitting data.

Advances in materials science and engineering played a crucial role in improving the durability and performance of spacecraft, enabling them to withstand the hostile environment beyond Earth's atmosphere. Another key aspect of technological innovation was the refinement of propulsion systems. Both nations engaged in a relentless quest for more powerful and efficient propulsion technologies to propel their spacecraft into orbit and beyond. This quest saw the emergence of pioneering solutions such as liquid fuel engines, powder thrusters and revolutionary propulsion concepts that laid the foundations for future space missions.

Advances in computer technology and software have played a vital role in the trajectory of space exploration. The development of sophisticated computer systems and software has made it possible to calculate precise trajectories, plan complex missions, and monitor space missions in real time. These advances have facilitated complex manoeuvres, deep space navigation, and the collection of large amounts of scientific data on celestial bodies. The proliferation of technological innovations has also transcended national boundaries. International collaborations and partnerships have emerged due to shared goals in space exploration. Notable examples include joint ventures for spacecraft development, collaborative research initiatives, and the exchange of knowledge and expertise in advanced aerospace technologies. These partnerships have not only accelerated the pace of technological progress, but have also fostered a spirit of cooperation and unity in the pursuit of space exploration. In short, the period marked by the space race between the United States and the USSR was the scene of an unprecedented wave of technological innovation that had repercussions in all fields of science and engineering. Embodying human ingenuity, these advances propelled humanity to a new frontier, laying the groundwork for the extraordinary achievements and discoveries that would shape the course of space exploration for years to come.

Key figures in rocket development

The progress of rocket development is inextricably linked

to the visionary individuals who dedicated their lives to advancing this cutting-edge field of technology. Countless personalities have left an indelible mark on the history of rocketry, propelling humanity towards the stars through relentless innovation and unwavering determination. Among these luminaries is the pioneering figure of Konstantin Tsiolkovsky, a Russian scientist whose theoretical work laid the foundations for modern astronautics. Tsiolkovsky's revolutionary research on rocket propulsion and space travel fundamentally shaped the trajectory of space exploration, inspiring generations of scientists and engineers to push the boundaries of what is possible.

Wernher von Braun, the eminent German aerospace engineer, also occupies a central place in the annals of rocket development. His essential contributions to the advancement of rocket technology, including the development of the V-2 rocket during World War II and his subsequent work with NASA, cemented his status as a preeminent figure in the field. Von Braun's expertise and leadership helped shape the trajectory of space exploration, leading to monumental achievements such as the Apollo moon landings. Sergei Korolev, the brilliant chief designer of the Soviet space programme, is another luminary in the field of rocket development. Renowned for his exceptional technical prowess and strategic vision, Korolev spearheaded the development of many pioneering spacecraft, including the world's first artificial satellite, Sputnik 1, and the iconic Vostok and Soyuz manned missions. His relentless pursuit of technological excellence and unwavering commitment to pushing the boundaries of space exploration cemented his legacy as a transformative force in rocket development.

The innovative contributions of Robert H. Goddard, nicknamed the 'father of the modern rocket,' had a profound impact on the evolution of rocket technology. Mr. Goddard's fundamental research in the field of liquid-fuelled rockets paved the way for significant advances in propulsion systems, laying the foundation for humanity's launch into space. His pioneering spirit and inventions left a lasting mark on the history of rocket development, demonstrating the power of visionary thinking and tireless dedication. These remarkable individuals, among many others, revolutionised the landscape of rocket development through their ingenuity, passion, and unwavering commitment to pushing the boundaries of human achievement. Their collective contributions form the cornerstone of modern space exploration, inspiring future generations to continue the quest for new frontiers and astronomical discoveries.

Major launches: Milestones

The history of rocket technology has been marked by many decisive moments, with major launches representing indelible milestones in the advancement of space exploration. Each launch represents the culmination of years of scientific and technical effort, propelling humanity towards new frontiers of knowledge and discovery. One of the most iconic launches that captivated the entire world was the historic Apollo 11 mission in 1969. This revolutionary achievement enabled man to set foot on the lunar surface for the first time, marking an unprecedented advance in our understanding of extraterrestrial environments. The images and words transmitted from the surface of the moon by astronauts

Neil Armstrong and Buzz Aldrin remain etched in the collective consciousness of humanity, symbolising the triumph of human ingenuity and determination. Another remarkable milestone was the launch of the Hubble Space Telescope in 1990, which revolutionised our understanding of the cosmos. Positioned in low Earth orbit, the Hubble telescope has provided unprecedented images of distant galaxies, nebulae, and other celestial phenomena, yielding invaluable insights into the origins and evolution of the universe. Its contributions continue to shape our understanding of astrophysics and cosmology, underscoring the profound impact of space observatories. More recently, Mars rover missions have embodied the spirit of exploration and technological innovation. Iconic missions such as Curiosity and Perseverance have enabled humanity to extend its reach to the Red Planet, conducting in-depth scientific research and paving the way for possible future human expeditions. The data and images transmitted back to Earth have shed light on the Martian landscape and provided tantalising clues about the planet's geological history and potential for habitability.

The successful deployment of international space stations, such as the International Space Station (ISS), is a testament to the successes of collaboration in the field of space exploration. The continued presence of astronauts conducting research and experiments in the microgravity environment of the ISS has led to invaluable scientific advances in various fields, including biology, materials science, and medicine. These initiatives demonstrate the considerable benefits that can be gained from international cooperation and coordination in space missions. Beyond these specific examples, countless satellite launches, orbital insertions, and deep space probes have expanded our understanding of the solar system and beyond. From the Voyager missions venturing into interstellar space to the New Horizons mission offering unprecedented views of Pluto, each launch represents a triumph of human curiosity and the quest for knowledge. When we reflect on these major launches and the milestones they represent, it becomes clear that space exploration continues to unite humanity in the pursuit of transcendent goals. These achievements serve as beacons of inspiration, highlighting the unlimited potential of scientific endeavour and the enduring spirit of exploration. As we look to the future, these achievements inspire us to go further, dream bigger, and embrace the challenges and opportunities that lie beyond our planet.

Global Impact of Rocket Technology

Rocket technology, with its profound ability to propel humanity beyond the confines of our planet, has had a tremendous global impact. The advent of rockets and space exploration has not only revolutionised our understanding of the universe, but has also exerted a significant influence on various aspects of life on Earth. One of the most striking impacts of rocket technology is its role in promoting international collaboration and competition. The space race between the United States and the Soviet Union in the mid-20th century not only spurred incredible scientific and technological advances, but also fuelled national pride and identity. This fierce competition transformed the political landscape and motivated unprecedented investments in education, research, and innovation on a global scale.

Rocket technology has played a key role in promoting international cooperation for the peaceful exploration and use of outer space. Collaborative missions, such as the International Space Station, have brought nations together around common goals, transcending geopolitical boundaries and fostering goodwill between countries.

The global impact of rocket technology extends to economic, industrial and societal dimensions. The development and application of rocketry has led to the emergence of new industries, employment opportunities, and economic growth, particularly through advances in satellite communications, navigation systems, and Earth observation technologies. These innovations play a vital role in addressing various global challenges, including climate change, disaster management, and resource monitoring.

Space exploration has inspired countless individuals to pursue careers in science, engineering, and technology, contributing to the formation of a highly skilled and specialised workforce that drives innovation across various sectors.

The cultural and inspirational impact of space exploration cannot be overstated. Iconic moments, such as man's first steps on the Moon, have captured the imagination of people around the world, sparking aspirations and ambitions that transcend cultural, linguistic and national boundaries. These monumental achievements are enduring symbols of human ingenuity and determination, inspiring future generations to dream big and reach for the stars. In short, the global impact of rocket technology spans domains ranging from geopolitics and international cooperation to economics, society, and culture. As we continue to push the boundaries of space exploration and technological innovation, it is essential to recognise and harness the vast implications of rocket technology to shape a prosperous and harmonious global future.

Continuing Evolution and the Modern Space Age

The ongoing evolution of rocket technology has propelled humanity into a new era of space exploration marked by unprecedented achievements and ambitious projects. As we enter the modern space age, it is becoming clear that advances in propulsion systems, spacecraft design, and international collaboration efforts have greatly expanded our understanding of the cosmos and redefined our ability to venture beyond the confines of Earth. One of the defining aspects of the modern space age is the diversification of space agencies and the emergence of private spaceflight companies, each contributing to a dynamic landscape of innovation and enterprise. With the rise of partnerships between government entities and commercial enterprises, space exploration and utilisation have taken on new dimensions, with tremendous implications for scientific research, economic growth, and global collaboration.

The advent of advanced propulsion technologies, such as ion engines and advanced chemical propulsion systems, has paved the way for sustainable missions to deep space and the prospect of interplanetary travel. These technological advances are paving the way for an era of extended space exploration, allowing us to reach distant celestial bodies with unprecedented precision and efficiency.

The modern space age has been marked by growing interest in space tourism and the rapid development of reusable launch systems, signalling a potential democratisation of access to space. This evolution towards sustainable and cost-effective space transport systems could revolutionise our relationship with space, making it more accessible to a wider spectrum of society and ushering in an era of unprecedented human presence beyond Earth's orbit. Looking ahead, the modern space age offers compelling opportunities to push our scientific boundaries, establish permanent human outposts beyond Earth, and harness the unlimited resources of the cosmos in a responsible and sustainable manner. The gradual convergence of innovative propulsion systems, advanced materials science, and visionary mission architectures promises transformative advances in space exploration and exploitation, marking an era defined by bold aspirations and unparalleled discoveries.

5
Triumphs and Trials
The Wonders of Nazi Engineering

Overview of Nazi technological advances

The technological advances made by the Nazis during the Second World War represented a significant leap forward in engineering and innovation. The Nazis placed an emphasis on research and development, which led to revolutionary advances in various areas of technology. One of the most advanced areas was that of weaponry and military vehicles. German engineers were responsible for the design and production of formidable tanks, such as the fearsome Tiger and Panther, whose firepower and armour were superior to those of their Allied counterparts. These advances revolutionised armoured warfare and influenced tank design for decades to come.

The Nazis made remarkable progress in aerospace technology by developing cutting-edge jet aircraft. The Messerschmitt Me 262, the world's first operational jet fighter, symbolises the advanced achievements of German aeronautical engineers. This innovative aircraft surpassed conventional propeller planes, highlighting the Nazi regime's prowess in aerospace engineering.

Nazi engineers played a vital role in the evolution of rocket and missile technology. The V-2 rocket, one of the first ballistic missiles, represents a monumental achievement in the field of propulsion and long-range weapons. Led by the brilliant mind of Wernher von Braun, this engineering marvel heralded the dawn of space exploration and served as a precursor to the advanced missile systems developed in the post-war period. In addition to these military advances, the

Nazis also ventured into other areas of technology, including communications, infrastructure, and medicine. Their advances in cryptography and communications technology not only supported their wartime efforts, but also laid the foundation for modern encryption methods.

The Autobahn, a pioneering motorway network, exemplifies the Nazis' forward-thinking approach to infrastructure development and transport engineering. Advances in medical research and human experimentation, while ethically controversial, yielded valuable knowledge in areas such as aviation medicine and trauma surgery. Although morally reprehensible, these discoveries contributed to the advancement of medical science and treatment methods. It is essential to examine the technological advances made by Nazi engineers to understand the complex legacy of innovations from the Second World War era. These advances have left an indelible mark on history, influencing subsequent technological progress and serving as a compelling reminder of the complex intersection between ethics, innovation, and the pursuit of scientific knowledge.

Technical breakthroughs in land vehicles

During the tumultuous period of the Second World War, Nazi Germany's technical breakthroughs in land vehicles demonstrated its remarkable technological prowess. The innovation and mastery of design in these vehicles revolutionised military operations and had a profound impact on the outcome of crucial battles. The development of the Tiger I and Tiger II tanks, renowned for their formidable armour and firepower, was a major advance. These heavy tanks represented a leap forward in armoured warfare and posed a formidable challenge to Allied forces on the battlefield.

The use of advanced suspension systems and powerful engines improved the mobility and manoeuvrability of these vehicles, allowing them to traverse difficult terrain with unprecedented agility. The Panther tank also proved to be a revolutionary engineering marvel, incorporating innovative sloped armour designs for improved protection and setting new standards for tank warfare.

The development of self-propelled guns such as the formidable Elefant and Jagdpanther demonstrated German engineering's prowess in integrating heavy weapons onto mobile platforms, thereby reshaping the dynamics of armoured combat. Furthermore, the iconic half-tracks, exemplified by models such as the Sd.Kfz. 251, demonstrated the versatility of German engineering, combining the capabilities of a tracked vehicle with the transport capacity of a truck, thereby enhancing the mobility and support capabilities of infantry units. These advances in land vehicle engineering not only reflect the technical acumen of German engineers, but also underscore strategic foresight in transforming military doctrine and tactics. Nazi Germany's relentless pursuit of technical excellence in land vehicles embodied a fervent desire to dominate the battlefield, setting a benchmark for future innovations in armoured warfare. As we delve into the intricate details of these technical triumphs, it becomes evident that the legacy of Nazi technological advances in land vehicles reverberates through the annals of military history, compelling us to critically examine their lasting impact and the lessons they impart.

Development of Strategic Infrastructure

Amid the chaos of World War II, Nazi Germany undertook an ambitious programme of strategic infrastructure development, aiming to establish a network capable of supporting its military operations and logistical needs. A comprehensive plan for the construction and maintenance of essential infrastructure, such as roads, railways, and bridges, was key to this endeavour. This infrastructure was designed to enable the rapid movement of troops, equipment and resources over vast distances, thereby improving the strategic mobility of the German military machine. The construction of autobahns, for example, not only facilitated the rapid deployment of forces, but also contributed to the economic revitalisation of the country. The concept of industrial self-sufficiency was central to the development of strategic infrastructure. The Nazis sought to build a strong industrial base capable of meeting the demands of war, thereby reducing their dependence on foreign imports. This desire resulted in the creation and expansion of factories, production facilities, and industrial zones strategically positioned to support the war effort. The integration of industrial infrastructure with military logistics played a critical role in the outcome of major battles and campaigns.

The Nazis used advanced engineering techniques to fortify defensive positions and create impregnable strongholds. The construction of complex defensive structures, such as the Atlantic Wall, demonstrates their commitment to fortifying key territories against potential invaders. These strategic infrastructure developments had a profound impact on the conduct of the war, effectively shaping the battlefield and influencing tactical decision-making.

The use of underground manufacturing facilities, such as the Mittelwerk complex, illustrated the Nazis' innovative approach to protecting critical assets from aerial bombardment. However, despite remarkable advances in infrastructure development, the ever-increasing demands of war placed a strain on available resources and manpower. As the war progressed, it became increasingly difficult to maintain and repair vital infrastructure due to material shortages and mounting pressure from the Allies. The destruction of key transport hubs and essential infrastructure by enemy bombing further exacerbated these difficulties, severely hampering German military operations.

In sum, the strategic infrastructure developments undertaken by Nazi Germany during the Second World War reflect a multifaceted approach encompassing transport, industry, defence and resource management. These efforts highlighted the link between military strategy, technical prowess and societal resilience, leaving a lasting legacy in the annals of history.

Innovations in weapons manufacturing

The period leading up to and during the Second World War was marked by a surge of innovation in the weapons manufacturing sector, particularly with the marvels of Nazi engineering. This era marked an important turning point in the application of scientific advances and industrial capabilities, resulting in the development of formidable weaponry that would redefine the nature of warfare. One of the main areas of interest was the creation of advanced firearms and weaponry designed to improve combat effectiveness. German engineers revolutionised small arms by producing iconic weapons such as the MP40 submachine gun, renowned for its reliability and performance on the battlefield.

The introduction of the StG 44 assault rifle represented an evolutionary leap in infantry firepower, laying the foundation for modern assault rifles. The integration of high-precision manufacturing techniques and innovative weapon configurations underscored the technological superiority of the German arms industry.

The strategic deployment of heavy artillery and anti-tank weaponry highlighted the prowess of the Nazi arms industry. The development of powerful tank guns, including the formidable 88 mm Flak cannon, demonstrated the ability to design powerful and versatile munitions capable of changing the course of battles. Beyond conventional weaponry, the era was also marked by an ambitious pursuit of advanced military technology, exemplified by the V-2 rocket—a remarkable achievement in long-range ballistic missile technology. The successful implementation of this weapon system marked the beginning of a new era in warfare, foreshadowing the future of missile development and space exploration. However, these advances in weapons manufacturing were not without ethical implications, as the use of scientific progress for destructive purposes raised profound moral dilemmas. While Nazi Germany's technical prowess left an indelible mark on the history of warfare, it also highlights the complex intersection of technological innovation and moral responsibility, prompting critical reflection on the impact of scientific achievements in times of conflict.

The role of engineers in the war effort

Engineers played a vital role in the outcome of the Second World War through their innovative designs, strategic planning, and technical expertise. Tasked with developing advanced weapons, infrastructure, and technologies, engineers were at the forefront of the war effort, contributing significantly to the success of their respective nations. The demands of war pushed engineering teams to push the boundaries of what was possible, leading to remarkable developments that continue to influence modern technology and industry. One of the key areas where engineers had a profound impact was in the design and production of military aircraft, including bombers, fighters, and reconnaissance planes. Their expertise in aerodynamics, propulsion systems and materials science led to the creation of advanced aircraft that redefined the nature of aerial warfare. These advances enabled strategic bombing campaigns, aerial reconnaissance missions and battles for air superiority that proved decisive in various theatres of war.

Engineers played a decisive role in the development of tanks and other armoured vehicles, which revolutionised ground combat. Leveraging their knowledge of mechanical engineering and automotive technologies, they designed and perfected armoured units that provided crucial support to the infantry and played an essential role in offensive and defensive operations. The ingenuity of engineers in creating more effective and powerful armoured vehicles had a direct impact on the dynamics of land warfare.

The critical role of engineers extended beyond traditional combat equipment to include the construction of vital infrastructure such as bridges, roads, and fortifications. Their expertise in civil and structural engineering enabled the rapid deployment of logistics networks, the reinforcement of defensive positions, and the facilitation of troop movements. The ability to rapidly construct and repair essential infrastructure proved indispensable in supporting the war effort and ensuring the mobility of military forces.

The application of innovative engineering solutions to the challenges of naval warfare highlighted the versatility and adaptability of engineering teams. From designing advanced ships to creating amphibious vehicles and landing craft, engineers demonstrated their ability to overcome complex maritime obstacles. Their contributions significantly influenced the success of naval operations and contributed to the overall strength of the Allied forces at sea.

In summary, the indispensable role of engineers in the war effort cannot be overstated. Their creativity, technical prowess, and unwavering dedication helped shape the course of World War II. Through their remarkable achievements, engineers not only supported military operations, but also laid the groundwork for the technological advances that would define the post-war era.

The Challenges of Resource Allocation and Management

During the tumultuous years of World War II, resource allocation and management posed formidable challenges to the wonders of Nazi engineering. Faced with a growing demand for materials and human capital to fuel its technological advances, the regime found itself confronted with a series of complex dilemmas that tested its leadership and logistical prowess. The need to efficiently allocate resources to a wide range of projects, from military equipment to infrastructure development, required meticulous planning and execution. One of the main challenges was the scarcity of critical resources such as steel, aluminium and fuel, as these products were essential for both military production and civilian subsistence. Competition for these resources often pitted different sectors of the war machine against each other, leading to complex power struggles and debates over priorities within the Nazi hierarchy.

Vast territorial conquests required the transport of resources over long distances, compounding logistical difficulties and making supply lines more vulnerable to enemy intervention. Human capital management also posed significant challenges. Skilled labour was in high demand for engineering, production, and the operation of advanced technologies. The regime struck a delicate balance between using its workforce for immediate military needs and investing in specialised training for long-term projects. At the same time, forced labour and the exploitation of prison populations raised ethical questions but were an important and controversial source of labour for the regime's ambitious projects.

Allocating resources to ensure sustainability amid Allied bombing and blockades required adaptive strategies. Civilian infrastructure was neglected as vital resources were diverted to military projects, creating social tensions and exacerbating the overall strain on the Nazi economy. The inefficiencies inherent in centralised control of resource allocation and management also hampered the regime's ability to respond dynamically to changing wartime conditions. Bureaucratic bottlenecks, miscommunication, and conflicting directives often led to wasted resources and missed opportunities for technological innovation. In sum, the challenges posed by resource allocation and management during the Second World War presented insurmountable obstacles to the wonders of Nazi engineering. The complex interplay between material scarcity, labour utilisation, ethical considerations, and strategic adaptability highlighted the immense difficulties faced by the regime in maintaining its technological ambitions amid the chaos of global conflict.

Ethical dilemmas and controversies

The pursuit of advanced engineering marvels during the Second World War gave rise to a myriad of ethical dilemmas and controversies that continue to be debated and reflected upon. One of the main ethical challenges faced by engineers and scientists working within the Nazi regime was the pervasive link between technological innovation and Nazi party ideology. The development and application of advanced technologies for military purposes raised profound moral questions about the responsibility of scientists and engineers in contributing to a destructive war. This prompted critical examination of the ethical limits of scientific activities, particularly when the results are used to perpetuate authoritarian agendas and human suffering. The controversial use of forced labour in the construction and maintenance of key infrastructure and armament facilities further compounded the ethical dilemmas. Engineers were confronted with complicity in the use of forced labour to advance technological dominance. The use of enslaved people in these projects left lasting scars on the conscience of those involved and sparked discussions about the responsibility of professionals in morally turbulent contexts.

The acquisition and integration of scientific knowledge derived from unethical and inhumane experimentation raised troubling ethical concerns. The involvement of certain individuals in inhumane medical experiments and the use of data obtained from these atrocities in technological advances cast a long shadow of moral unease. These troubling practices have highlighted the disturbing convergence of scientific progress and ethical transgressions, prompting contemporary reflections on the responsibilities of researchers and engineers in difficult ethical environments. Ethical dilemmas and controversies have extended beyond periods of war, shaping current discourse on the moral implications of technological progress. By delving into these complex and controversial dimensions, we are compelled to confront the multifaceted ethical challenges faced by engineers in times of global conflict and the enduring lessons that continue to resonate in contemporary society.

Comparative analysis with the efforts of Allied engineers

After the end of the Second World War, a critical analysis of the wonders of Nazi engineering compared to their Allied counterparts reveals important information about the technological landscape that shaped the outcome of the war. The comparative analysis of Nazi and Allied engineering efforts highlights the innovative advances made by both sides in areas such as land vehicles, aviation, weaponry, and strategic infrastructure during the global conflict.

From the outset, it is imperative to recognise the notable advances made by Nazi engineers alongside the advanced technologies developed by the Allied forces. The comparative analysis examines the overall impact of engineering dynamics on the progression and outcome of the Second World War. An examination of the design and deployment of land vehicles shows that both the Axis and Allied powers demonstrated exemplary expertise in engineering marvels such as tanks, armoured vehicles and transport units. The comparative study highlights the effectiveness and tactical advantages conferred by the different engineering philosophies employed by the opposing factions, illuminating the nuanced approach to mechanised warfare by considering factors such as mobility, firepower, and survivability.

The comparative assessment extends to the field of aviation, where innovative advances in aeronautical technology have significantly influenced the trajectory of aerial warfare. From the iconic Messerschmitt and Focke-Wulf of the Luftwaffe to the formidable American P-51 Mustang and British Spitfire, the competitive nature of technological advances in aerial combat illustrated the engineering prowess on both sides of the wartime divide.

In the field of armaments, comparative analysis reveals the sophisticated developments that marked the evolution of weapons and ammunition. It elucidates the strategic implications of advances in firearms design, artillery and ammunition production, highlighting the essential role played by the ingenuity of engineers in reshaping the dynamics of the battlefield and influencing the tactical doctrines adopted by military strategists. Strategic infrastructure, an often-overlooked facet of wartime engineering, emerges as a crucial area for comparative study. The construction of motorways, bridges and fortifications testifies to the technical prowess that underpinned logistical operations and defensive measures, thereby influencing the manoeuvrability and resilience of Axis and Allied forces in various theatres of conflict. Ultimately, comprehensive comparative analysis serves as an invaluable lens for discerning the profound technological impacts that underpinned the outcomes of the Second World War, highlighting the considerable influence of engineering marvels on the course of history.

Technological Impacts on the Outcomes of World War II

Technological advances during World War II played a critical role in shaping the outcome of the war. This progression extended to various fields, including weaponry, communication, transportation, and logistics. Rapid advances in military technology significantly influenced the strategies and tactics employed by the Allied and Axis powers, ultimately impacting the direction and resolution of many engagements. The development and deployment of superior weapon systems and equipment often conferred significant advantages on the belligerent forces, greatly influencing the trajectory of the conflict.

One of the most notable impacts of technology on the outcome of the Second World War is evident in the field of aerial warfare. The advent of long-range bombers and advanced fighter aircraft made it possible to achieve unprecedented range and accuracy in the conduct of air missions. These capabilities reshaped the landscape of warfare, altering the dynamics of strategic bombing campaigns and battles for air superiority.

Advances in radar technology provided crucial tactical advantages and greatly improved situational awareness, shifting the balance of power in aerial engagements. Furthermore, naval warfare underwent a paradigm shift thanks to technological innovations such as sonar systems, amphibious landing craft, and improved naval artillery. The implementation of these advances transformed naval operations, affecting maritime trade routes and fundamentally impacting the effectiveness of blockades and naval engagements.

Advances in submarine warfare and convoy protection had far-reaching repercussions, influencing control of vital sea lanes and disrupting enemy supply lines. The importance of technological developments was also evident on the ground, where armoured vehicles, artillery and infantry weapons saw substantial advances. Mechanised warfare, characterised by the integration of tanks, armoured vehicles and mobile artillery, changed the scope and scale of land engagements. The mechanisation of warfare facilitated rapid manoeuvres and redefined the dynamics of offensive and defensive operations, significantly influencing the outcome of decisive battles and campaigns.

In short, the technological impacts of the Second World War reverberated across the global theatre, profoundly affecting the strategies, tactics and final outcomes of the conflict. The use of advanced weapons, communications systems, and logistical infrastructure reshaped the battlefield, transforming the nature of warfare and significantly influencing the course of history. Understanding the multifaceted technological impacts provides insight into the complexities of warfare and highlights the enduring legacy of innovation and progress in the tumultuous context of global conflict.

Transition to the use of post-war technologies

At the end of the Second World War, the world witnessed a paradigm shift in innovation and the use of technology. The war served as a catalyst for rapid advances in various fields, fundamentally altering the global industrial landscape and ushering in an era of unprecedented progress and development. As nations sought to rebuild and restructure in the aftermath of the conflict, the integration of wartime technologies into civilian and commercial sectors became a focal point for post-war economies. The transition to the use of post-war technologies was marked by an influx of scientific and technical talent that had played a key role in military innovations during the war. These experts brought with them a wealth of knowledge and experience, ready to redirect their skills towards peacetime applications. This period was marked by a boom in research and development, as industries shifted away from war production to focus on creating consumer goods and infrastructure that could support a burgeoning global economy.

One of the most remarkable aspects of the post-war transition was the conversion of military technologies for civilian use. Innovations such as radar, jet engines and advanced materials found new applications in aviation, telecommunications and manufacturing processes. The harnessing of nuclear energy, which had been central to the war effort, now promised countless possibilities for energy production and scientific exploration. These examples illustrate the multifaceted impact of wartime technologies, which permeated every aspect of post-war society.

The spin-offs from military research and investment catalysed the birth of entirely new industries. The aerospace sector experienced unprecedented growth, stimulated by the accumulation of knowledge and capabilities developed during the war. Similarly, advances in computing and electronics, previously concentrated in military establishments, quickly found their way into commercial markets, laying the foundations for the digital revolution that would follow in the decades ahead. The process of transitioning to the use of post-war technologies was not without its difficulties. The conversion of existing facilities, the retraining of personnel, and the realignment of supply chains presented formidable obstacles to overcome.

The ethical implications of integrating wartime technologies into civilian contexts required careful consideration, particularly with regard to the responsible management and regulation of potentially disruptive advances. In retrospect, the post-war period stands as a testament to the resilience and adaptability of human ingenuity. The successful integration of wartime technologies for peaceful and productive purposes has been a lasting legacy of innovation, shaping the trajectory of technological progress and establishing a blueprint for the constructive exploitation of scientific advances for the benefit of humanity.

6
Jet Propulsion
Redefining Warfare And Aviation

Jet propulsion

Jet propulsion is a revolutionary method of propelling aircraft and other machines by expelling streams of gas or liquid at high speed. This revolutionary technology is based on the fundamental principle of Newton's third law of motion, which states that every action has an equal and opposite reaction. In the context of jet engines, this translates into the forward movement of an aircraft thanks to the reactive force produced by the expulsion of exhaust gases at high speed in the opposite direction. The mechanics of jet propulsion involve air intake, air compression, mixing with fuel, ignition, and then expulsion, creating a continuous cycle of thrust generation. Unlike traditional propeller systems, jet propulsion more efficiently harnesses the potential energy of fuel to produce immense power and speed. The functionality of jet propulsion is rooted in thermodynamics and the efficient conversion of chemical energy into kinetic energy. This process involves the careful combustion of fuel in the engine, causing a rapid increase in temperature and pressure, leading to the expulsion of exhaust gases at high speed. Understanding the complex interaction between airflow, combustion, and exhaust gas dynamics is essential to understanding the seamless operation of jet propulsion.

The evolution of jet propulsion has continually pushed the boundaries of engineering and aerodynamics, enabling unparalleled advances in the aviation industry and beyond. Through mastery of these principles, jet propulsion has redefined the possibilities of flight and influenced a myriad of industries, from the military and transport to space exploration and beyond.

Historical context: The emergence of jet engines

The development of jet engines represents a pivotal moment in the history of aviation and warfare, as it revolutionised the way we understand and utilise propulsion. The concept of jet propulsion dates back to the early 20th century, when aviation pioneers and engineers were actively exploring alternatives to traditional propeller-driven aircraft. However, it was not until the eve of the Second World War that significant progress was made in transforming the theoretical idea of jet propulsion into a tangible reality.

In fact, it was during this period that German and British engineers made considerable progress in the development of jet propulsion technology. German engineer Hans von Ohain and British engineer Frank Whittle were particularly instrumental in designing viable jet engines during this period. Their pioneering work laid the foundation for the future of aviation and military technology, catapulting the world into a new era of air power.

As the war progressed, both Germany and the Allied forces recognised the strategic implications of jet engines. The emergence of jet aircraft, such as the German Messerschmitt Me 262 and the British Gloster Meteor, demonstrated the profound impact of this technology on aerial combat. This evolution in propulsion technology not only redefined the capabilities and limitations of aircraft, but also influenced the outcome of crucial engagements during the war. The rapid development and deployment of these advanced aircraft marked a turning point in the evolution of aviation and warfare, paving the way for the dominance of jet propulsion after the war. The historical context surrounding the emergence of jet engines not only reflects a technological arms race, but also highlights the transformative power of innovation in shaping the trajectory of global conflicts and the future of flight.

Technological Breakthroughs and Innovations

Following the historic advent of jet engines, the field of aeronautical engineering underwent a radical transformation marked by unprecedented technological breakthroughs and innovations:

As the Second World War raged, the urgency to propel military capabilities led to profound advances in jet propulsion technology, forever changing the course of warfare and aviation. The pioneers of this extraordinary era relentlessly sought solutions to the formidable challenges posed by harnessing the power of jet engines for practical purposes. One such revolutionary innovation was the development of axial-flow compressors, which significantly improved engine performance, enabling higher thrust and more efficient fuel consumption. This breakthrough not only revolutionised aircraft propulsion, but also laid the foundation for further advances in gas turbine technology. The introduction of new materials and manufacturing techniques facilitated the construction of robust and lightweight engine components, ensuring unparalleled reliability and operational endurance. The redesigned airframes of jet aircraft were carefully engineered to withstand the intense forces generated by high-speed flight, ushering in a new era of aerodynamic design.

The integration of advanced electronic systems, such as improved instrumentation and radar technologies, enhanced the navigation and combat capabilities of jet aircraft. Innovations in propulsion systems, including afterburners and variable-area nozzles, further amplified the speed and agility of jet platforms, improving their strategic versatility in combat scenarios.

Refinements in aerodynamic air intake design have improved engine efficiency across a wide range of altitudes and speeds, heralding a paradigm shift in the performance envelope of military aircraft. These remarkable technological breakthroughs and innovations not only redefined the parameters of aviation, but also reshaped the trajectory of military operations, laying the foundation for an era of unprecedented air capabilities and strategic dominance.

Strategic Implementation During World War II

During the Second World War, the strategic implementation of jet propulsion technology played a critical role in reshaping the dynamics of aerial warfare. The introduction of jet aircraft offered significant advantages, including higher speeds and more efficient performance at high altitudes, revolutionising air combat tactics and capabilities. Germany's deployment of jet aircraft, such as the Messerschmitt Me 262, fundamentally changed the landscape of aerial warfare, posing a formidable threat to Allied forces. Whether intercepting enemy bombers or engaging in dogfights, these innovative aircraft posed new challenges for the Allies, who had to find effective ways to counter them. The impact of jet propulsion extended beyond aerial combat, with reconnaissance and ground attack missions benefiting from the increased speed and manoeuvrability of jet aircraft.

The psychological impact of encountering these advanced aircraft on the battlefield cannot be underestimated, as it caused concern and uncertainty among Allied pilots and ground personnel. On the Eastern Front, the Soviet Union also recognised the strategic importance of jet propulsion and began developing its own jet fighters in response to German advances. As the war progressed, the United States and its allies intensified their efforts to understand and exploit jet propulsion technology, recognising its transformative potential for future conflicts. This era marked a decisive turning point in aviation history, laying the foundation for the widespread adoption of military and commercial jet aircraft in the post-war period. The strategic implementation of jet propulsion during the Second World War not only influenced the outcome of specific engagements, but also paved the way for the dominance of jet aviation in the decades that followed.

The Impact of Jet Propulsion on Aircraft Design

The introduction of jet propulsion during the Second World War marked a monumental shift in aircraft design, revolutionising the principles and capabilities of aviation. Jet propulsion brought about a paradigm shift by providing aircraft with unprecedented speed, altitude, and manoeuvrability, thereby reshaping the entire landscape of aerial warfare. The primary impact of jet propulsion on aircraft design was the transition from propeller engines to jet engines, which fundamentally changed the way aircraft were built and operated. This change had profound implications for structural design, aerodynamics, and overall performance. The integration of jet propulsion required the development of advanced airframes capable of harnessing the immense power and speed generated by jet engines. As engineers grappled with new speeds and stresses, aerodynamic theories were re-evaluated and refined, leading to the emergence of sleeker, more streamlined aircraft shapes optimised for high-speed flight.

Increased thermal stresses and extreme operating conditions created a need for improved materials and construction techniques, which ultimately led to advances in aircraft manufacturing. In addition to technical considerations, aircraft designers also had to reconsider factors such as fuel efficiency, range, and payload capacity to fully exploit the advantages of jet propulsion. The transition to jet aircraft gave rise to innovations such as swept wings and delta wings, which helped overcome the aerodynamic challenges posed by transonic and supersonic flight. These developments improved performance and facilitated the exploration of new flight regimes previously inaccessible to conventional propeller aircraft.

Jet propulsion catalysed the evolution of avionics and control systems, requiring greater sophistication in navigation aids, fly-by-wire technology, and cockpit instrumentation to keep pace with the capabilities of jet aircraft. The impact of jet propulsion extended beyond military applications, influencing civil aviation and laying the foundation for a new era of commercial air transport. The adoption of jet propulsion in airliners accelerated travel times, expanded global connectivity, and improved passenger comfort.

The demand for more efficient and reliable engines has led to constant improvements in propulsion systems and engine technologies, further enhancing the safety and economy of air transport. In fact, jet propulsion marked a turning point in the history of aircraft design, catalysing a wave of innovations that propelled aviation into the modern era and continues to shape the trajectory of aerospace engineering.

Comparative Analysis: Axis and Allied Jet Technologies

With the emergence of jet propulsion during the Second World War, the Axis and Allied powers embarked on revolutionary developments in aviation technology. This era saw the birth of groundbreaking technologies that would alter the course of aviation history. A comparative analysis of the jet technologies implemented by the Axis and Allied forces reveals distinct approaches and their respective impacts on the trajectory of the war and aviation. The Axis powers, primarily Germany, spearheaded the deployment of advanced jet aircraft such as the Messerschmitt Me 262, the world's first operational jet fighter. German engineering prowess in jet propulsion represented a significant leap forward in aviation capabilities.

On the other hand, the Allied forces, particularly the United States and Britain, fervently pursued their own development of jet technologies, with notable achievements such as the Gloster Meteor and the pioneering work of the Whittle engine. The differing strategies of the Axis and Allied powers in jet propulsion had a considerable impact on the conduct of aerial warfare.

Here, we examine the technical nuances, strategic decisions, and operational significance of Axis and Allied technologies in the field of jet propulsion. By dissecting the design principles, performance, and combat applications of these respective advances, this comparative analysis aims to highlight the unique contributions and lasting legacies of Axis and Allied jet aircraft technologies.

Examining the logistical and industrial challenges faced by both sides sheds light on the different trajectories of technological progress and wartime utilisation. In this context, we examine how operational doctrine, resource allocation, and technological exchanges influenced the evolution of jet aircraft technologies on both sides. This comprehensive comparison underscores the central role of jet propulsion in redefining the dynamics of aerial warfare and shaping the future of aviation. Meticulous examination of Axis and Allied jet aircraft technologies illuminates the complex interplay between innovation, industrial capacity, and strategic imperatives. Such comparative analysis not only enriches our understanding of the historical development of jet propulsion, but also offers valuable insights into the enduring legacy and ongoing evolution of advanced aeronautical technologies.

Operational Challenges and Solutions

With the advent of jet propulsion, military powers faced a myriad of operational challenges when integrating this revolutionary technology into their arsenals. The rapid acceleration and altitude-gaining capabilities of jet aircraft gave rise to new tactical considerations, particularly with regard to fuel consumption, engine reliability, and maintenance requirements. The operation of jet engines posed significant technical challenges, such as managing high temperatures, compressibility effects, and turbine design.

The transition from piston-powered aircraft to jet aircraft required extensive training for pilots and ground personnel to adapt to the distinct flight characteristics and maintenance requirements. The limited availability of raw materials suitable for manufacturing advanced jet engines exacerbated logistical challenges during wartime. To overcome these considerable obstacles, engineers and aviation specialists relentlessly sought innovative solutions. They developed new cooling mechanisms and materials to withstand extreme temperatures, optimised fuel delivery systems to improve efficiency, and refined aerodynamic designs to maximise performance. This period of intensive problem-solving led to breakthroughs that shaped the evolution of jet propulsion. Collaborative efforts among Allied nations led to the standardisation of jet fuel types and facilitated the establishment of global maintenance and supply networks, providing sustainable operational support.

The establishment of training programmes and specialised manuals enabled personnel to acquire the expertise necessary to effectively operate and maintain jet aircraft. These strategic adjustments enhanced operational readiness and streamlined the integration of jet aircraft into combat missions. After the war, the experience gained in overcoming these challenges led to continuous improvements in jet engine technology and paved the way for civilian applications. The rapid expansion of commercial air transport saw the successful adaptation of jet propulsion for passenger aircraft, revolutionising global transportation. The multifaceted operational challenges encountered during the early adoption of jet propulsion not only accelerated technological innovations but also highlighted the resilience and ingenuity of the human intellect.

Legacy and post-war evolution

After the end of the Second World War, the legacy of jet propulsion technology reverberated across countries, leaving an indelible mark on the trajectory of aviation and warfare. The end of the war marked a paradigm shift in the use of jet engines, with countries seeking to harness this revolutionary technology for civilian purposes and military supremacy. Here, we examine the lasting impact and evolution of jet propulsion after the war, highlighting its transformative influence on global aeronautics. As the spectre of conflict faded, the remarkable potential of jet propulsion was unveiled in the commercial sphere, ushering in a new era of air travel. As airlines capitalised on the superior speed and efficiency offered by jet engines, a wave of innovation swept through the industry, propelling aircraft design and performance to unprecedented levels. The integration of jet propulsion revolutionised commercial aviation, ushering in an era of supersonic travel and expanded global connectivity. At the same time, military forces continued to refine and advance jet propulsion technology, leading to the proliferation of advanced aircraft capable of executing strategic missions with unmatched precision and speed. The post-war period saw the convergence of civil and military applications of jet propulsion, giving rise to multifaceted breakthroughs that transcended traditional boundaries.

The rapid evolution of jet propulsion stimulated collaborative efforts between governments, aerospace engineers, and manufacturers, fostering a climate of innovation and competition that accelerated technological progress. This phase of rapid development laid the foundation for the prolific advances in aviation and propulsion systems that define contemporary aerospace engineering. Beyond its immediate impact, the legacy of jet propulsion after the Second World War underscores the enduring importance of technological adaptation and ingenuity in shaping the course of modern history. The ongoing evolution of jet engines stands as a testament to human innovation and resilience, serving as a poignant reminder of the profound transformations that can emerge from conflict and adversity.

The Transition to Commercial Aviation

At the end of the Second World War, the aviation industry underwent a monumental shift from military applications to commercial endeavours. The rapid evolution of jet propulsion during the war opened up new possibilities for civil aviation and ushered in a wave of innovation that would redefine air transport. This crucial transition not only shaped the trajectory of commercial aviation, but also revolutionised global connectivity and transport. With surplus warplanes being repurposed for civilian use, demand for efficient and reliable air transport skyrocketed. The post-war period saw an influx of advances in aircraft design, manufacturing processes, and operating techniques to meet the growing demand for commercial air travel. Airlines embraced the potential of jet aircraft, recognising their ability to offer unprecedented speed, range and comfort to passengers. This led to the introduction of iconic airliners such as the Boeing 707 and Douglas DC-8, marking a significant break from the era of propeller-driven aircraft.

The transition to commercial aviation gave rise to new considerations regarding safety, infrastructure, and regulatory frameworks. The need to standardise air traffic control systems, airport facilities, and international regulations became increasingly imperative as commercial airlines expanded their routes across continents. Governments and industry stakeholders collaborated to establish guidelines and protocols that would ensure the safety and efficiency of commercial air transport.

As air travel grew in popularity with the public, the emergence of commercial aviation not only democratised access to distant destinations, but also stimulated economic growth and cultural exchange on a global scale. The introduction of transatlantic and transpacific flights reduced travel times and redefined the concept of long-distance travel, profoundly impacting tourism, commerce, and interpersonal relationships. Essentially, the transition to commercial aviation propelled the world into a new era of interconnectedness, accessibility, and mobility. The convergence of technological advances, market dynamics and societal aspirations paved the way for transformative change in how people perceive, experience and utilise air travel. This paradigm shift continues to resonate today, shaping the contemporary landscape of commercial aviation and demonstrating the enduring legacy of wartime jet propulsion.

Future Outlook: Next-Generation Jet Technologies

As we look to the future, the aviation world is poised for remarkable advances in jet propulsion technologies. The advent of next-generation jet propulsion systems promises to revolutionise air travel, military operations and aerospace engineering as a whole. Through continuous research and development, there is a clear path toward more efficient, environmentally friendly, and higher-performing jet engines. One of the most significant breakthroughs is in the field of supersonic and hypersonic flight. Engineers and scientists are at the forefront of developing innovative propulsion systems capable of exceeding the speed of sound with increased safety and efficiency. By leveraging advanced materials and cutting-edge design principles, the next generation of supersonic and hypersonic jets aims to redefine global air travel, offering unprecedented speed and connectivity.

The integration of electric propulsion and hybrid engine technologies represents another decisive advance in the aviation landscape. These futuristic propulsion systems are poised to mitigate the environmental impact of air travel by reducing emissions and noise pollution. From all-electric aircraft to hybrid jets, the convergence of sustainability and aviation is driving the evolution of next-generation jet technologies. This transformative shift not only aligns with global sustainability initiatives, but also ushers in a new era of environmentally friendly and economically viable air travel.

Artificial intelligence (AI) and autonomous systems are set to play an important role in the future of jet aircraft technology. AI-driven flight control systems and autonomous optimisation algorithms are being developed to improve aircraft performance, safety and operational efficiency. Through the seamless integration of AI, next-generation jets are expected to offer unparalleled levels of reliability, adaptive responsiveness and advanced self-diagnostic capabilities, raising the bar for safety and operational excellence in the aviation industry. In addition to these advances, the convergence of additive manufacturing and 3D printing techniques is driving a paradigm shift in the production of jet engine components.

This change promises complex designs, reduced material waste and improved structural integrity, propelling the development of lighter and more durable jet engines. The application of these advanced manufacturing methodologies is expected to accelerate production processes and promote greater flexibility in customising jet propulsion systems to specific performance requirements.

In short, the future of jet propulsion is extremely promising, thanks to the convergence of revolutionary technologies and visionary innovations. As the frontiers of aviation continue to expand, next-generation jet technologies are poised to usher in an era of unprecedented speed, sustainability, and safety. With each step forward, aviation enthusiasts, engineers and industry leaders await the dawn of a new era, a future where the sky will be synonymous with progress, potential and unparalleled possibilities.

7
The Pursuit Of The Allies

Strategic objectives and initial movements

The early stages of the Allied pursuit were marked by a meticulous examination of geopolitical objectives and military strategies, aimed at decisively changing the course of the war. The main strategic objective was to dismantle the military power of the Axis powers, particularly Nazi Germany, and secure an advantageous position for the Allies. This required coordinated offensives on multiple fronts, with an emphasis on air supremacy, amphibious assaults, and land invasions. The initial move involved the implementation of Operation Overlord, the largest maritime invasion in history, which aimed to gain a foothold in Normandy and provide a launching pad for the liberation of Western Europe.

At the same time, operations in the Pacific theatre aimed to repel Japanese forces and regain control of strategic territories. These simultaneous movements underscored the global scale and complexity of the Allied pursuit, which required precise coordination and unwavering determination. The use of various military tactics, including aerial bombardment, clandestine operations, and infantry combat, highlighted the comprehensive approach taken to achieve key strategic objectives. As the conflict unfolded, the adaptation of strategies to changing circumstances and intelligence information played a critical role in the trajectory of the Allied pursuit. The complex interaction between military command, strategic planning, and operational execution significantly influenced the outcome of major battles and campaigns. The ability to exploit technological advances, such as radar and encryption technologies, notably increased the effectiveness of the Allied pursuit. The multifaceted nature of the strategic objectives and initial movements reflects the seriousness of the Allied pursuit and the unified determination to change the course of history.

Key figures leading the Allied effort

To counter the formidable technological advances of the Axis powers, key figures played a decisive role in coordinating the Allied effort. These individuals demonstrated unwavering determination, keen strategic insight, and a deep understanding of the stakes involved in the race for advanced technology and its exploitation for the greater good. Among these influential figures, General Leslie Groves played a vital role in overseeing the Manhattan Project, directing the scientific and industrial effort that culminated in the manufacture of the first atomic bombs. His resolute leadership and management skills were essential in navigating the complexities of this ambitious undertaking, propelling the Allied powers to a significant technological advantage. At the same time, the tenacious Winston Churchill, with his resounding rhetoric and profound knowledge of international affairs, provided unwavering support and direction, rallying the spirit of cooperation among the Allies.

The innovative vision and determination of scientist Vannevar Bush, who headed the Office of Scientific Research and Development, enabled crucial advances in radar technology, scientific research, and military medicine, thereby strengthening the Allies' arsenal. Complementing the aforementioned leaders, Admiral William D. Leahy, Chief of Staff to the President, exerted tremendous influence by promoting cohesion and synchronisation among the various branches of the US military, laying the foundation for effective collaboration.

The wise leadership of Sir Henry Tizard, a brilliant scientist and advisor to the British government, played a key role in facilitating the exchange of technological knowledge and encouraging innovation across borders. These figures, among others, demonstrated remarkable foresight and determination, navigating the complex web of scientific, political and military domains and significantly shaping the course of history through their leadership in the pursuit of technological superiority.

Coordination among the Allied Powers

During the tumultuous years of World War II, coordination among the Allied Powers was a key factor in determining the outcome of the conflict. Effective collaboration between nations such as the United States, the United Kingdom, the Soviet Union, and other Allied forces played a vital role in the development of military strategies, intelligence operations, and technological advances. This collaboration required a delicate balance of diplomatic negotiations, military planning, and resource allocation to form a unified front against the Axis powers.

The organisation of high-level conferences, such as the Tehran Conference, the Yalta Conference and the Potsdam Conference, allowed the leaders of the Allied powers to discuss and coordinate their military and political strategies. These meetings enabled influential figures such as Winston Churchill, Franklin D. Roosevelt and Joseph Stalin to align their objectives and determine the direction of the war effort. These conferences resulted in agreements on post-war occupation zones, war crimes trials, and the formation of international organisations, paving the way for post-war stability and cooperation.

In addition to diplomatic efforts, military coordination among Allied forces involved complex planning, joint operations, and the exchange of crucial intelligence. The integration of military commands, such as the combined chiefs of staff, facilitated the synchronisation of military campaigns and the pooling of resources toward common goals. The successful execution of large-scale operations, including the Normandy landings and offensives on the Eastern Front, demonstrated the effectiveness of coordinated military efforts in achieving significant victories over the Axis powers.

It is important to recognise the contributions of Allied intelligence agencies, which facilitated coordination through the collection, analysis, and dissemination of vital information. The exchange of intelligence between organisations such as the Office of Strategic Services (OSS), the British Secret Intelligence Service (SIS), and the Soviet Main Intelligence Directorate (GRU) played a crucial role in providing decision-makers with actionable information and assessments of enemy capabilities and intentions. Ultimately, coordination among the Allied powers demonstrated the strength of unity in the face of adversity. By aligning their political, military, and intelligence efforts, the Allied nations effectively leveraged their collective power to meet the challenges posed by the Axis powers.

Intelligence-gathering operations

During World War II, intelligence-gathering operations played a vital role in the Allies' pursuit of Nazi technology and scientific advances. The relentless quest for valuable information required sophisticated and covert strategies, often carried out by specialised units and agents in different theatres of war. A notable aspect of these operations was the use of human intelligence (HUMINT) and technical intelligence (TECHINT), each serving distinct but complementary purposes. HUMINT operations involved skilled agents infiltrating enemy territory, cultivating informants, and executing risky reconnaissance missions to gain first-hand knowledge of enemy technological developments. These daring individuals operated in the utmost secrecy and faced grave dangers, exemplifying the courage and dedication that are at the heart of the intelligence community's efforts. Their reports provided invaluable information on the enemy's scientific breakthroughs, industrial capabilities, and key personnel, thus constituting essential elements for strategic decision-making within the Allied command structure.

At the same time, TECHINT efforts focused on capturing and analysing tangible artefacts of technological innovation, whether captured equipment and documents, intercepted communications, or enemy research facilities. Specialised teams of experts meticulously examined these documents, extracting vital intelligence on the nature and extent of the enemy's technological achievements. This technical intelligence not only demystified the enigmatic field of advanced weaponry and scientific experiments, but also revealed the vulnerabilities and limitations of the Axis powers' innovations. The symbiotic relationship between HUMINT and TECHINT operations facilitated a comprehensive understanding of the enemy's technological landscape, enabling the Allies to strategically direct their pursuits and counter potential threats posed by Nazi advances.

Collaboration between the intelligence services of different Allied nations fostered a collective approach to gathering and synthesising vital information, highlighting the importance of international cooperation in pursuing common goals. Intelligence-gathering operations were not without perilous challenges and ethical dilemmas. The clandestine nature of these activities required meticulous risk assessment, operational security, and adherence to strict ethical guidelines to protect the lives of agents and mitigate the potential repercussions of intelligence leaks.

The effective use of the intelligence gathered required a delicate balance between exploiting enemy innovations and respecting moral and legal considerations, reflecting the complex interaction between war, morality, and technological progress. Ultimately, the intelligence gathered during these meticulous operations greatly influenced the trajectory of the war and post-war developments, shaping the course of technological innovation and geopolitical dynamics in the decades that followed.

Technological rivalry and captured innovations

At the height of the Second World War, technological rivalry reached its peak, with the Allied forces actively seeking to capture and exploit Nazi innovations. The race for the enemy's cutting-edge technologies became an essential part of the war effort, with both sides vying for scientific breakthroughs that could tip the balance in their favour. This fervent quest for captured innovations played out on various fronts, ranging from industrial espionage and clandestine operations to daring raids and strategic warfare. The Allies recognised the potential of acquiring advanced German weapons, aerospace designs and revolutionary scientific discoveries to bolster their own arsenals and push the boundaries of military capabilities. The fierce competition to seize these innovations resulted in a remarkable display of ingenuity and determination on the part of Allied scientists, engineers, and agents, who worked tirelessly to analyse, replicate, and integrate the captured technologies into their own war machines. This frenzied technological rivalry not only enabled the Allied forces to make numerous advances, but also fostered an unprecedented acceleration of scientific progress that reverberated far beyond the battlefield. The captured innovations played a decisive role in the trajectory of post-war innovation and technological development, influencing fields such as aviation, rocketry, medicine and computing.

The secret acquisition of enemy technologies raised complex moral and ethical dilemmas, prompting deep introspection among the Allies about the implications of exploiting wartime discoveries. The resulting debates and decisions regarding the use of captured innovations highlighted the heavy responsibilities incumbent upon the victors when it came to utilising new scientific achievements for both military and civilian purposes. The struggle for supremacy in wartime technological prowess is an essential chapter in the annals of history, leaving an indelible mark on the evolution of science, industry, and global geopolitics.

Ethical considerations of technological seizure

The pursuit and seizure of technological innovations during times of conflict raise profound ethical considerations that continue to resonate throughout history. As the Allied powers strove to gain an advantage over their adversaries, the quest for technological superiority placed them at a moral crossroads. The pressing need to acquire advanced enemy technologies for defensive and offensive purposes confronted decision-makers with weighty ethical dilemmas. Here, we note the importance of the complex web of ethical considerations surrounding the capture and use of enemy innovations. One of the primary ethical considerations is the question of appropriation versus theft. The line between justified acquisition and outright theft of technology from defeated adversaries has been the subject of intense debate. A delicate balance has had to be struck between the imperative to accelerate technological progress and the need to respect ethical standards of fair play and intellectual property rights.

The ethical implications of exploiting innovations captured from the enemy extend to the use of knowledge derived from research and development conducted under oppressive or inhumane conditions. The potential dilemma of benefiting from advances born of human suffering and unethical practices presents a formidable moral dilemma. Another ethical dimension concerns the responsible treatment of captured technologies to avoid unintended harm or misuse. This involves ensuring that acquired innovations are used for noble purposes and do not perpetuate further conflict or injustice.

Ethical considerations surrounding the seizure of technology include the treatment of scientists and technicians involved in the development of seized technologies. This involves issues such as their rights, welfare, and the extent to which their expertise should be used in the scientific endeavours of the nation seizing the technology. It also raises questions about accountability and the pursuit of justice for individuals complicit in unethical research and production during wartime. Reflecting on these ethical dimensions serves as a reminder of the ongoing relevance of ethical considerations in the context of war and technological progress. In navigating the complexities of technological seizures, the Allied powers were forced to confront and grapple with the ethical dimensions of their actions, leaving an indelible mark on the intersection of war, morality, and innovation.

Key encounters and turning points

As the conflict raged, major encounters and turning points marked important milestones in the pursuit of technological advances and strategic victories. One such crucial moment was the interception and decryption of vital enemy communications, which provided invaluable information about the adversary's plans and operations. This breakthrough not only enabled the Allies to anticipate and counter enemy tactics, but also revealed the extent of their technological capabilities, informing subsequent efforts to neutralise and exploit captured innovations.

Notable encounters on the ground demonstrated the adaptability and resilience of Allied forces in the face of evolving threats, highlighting the effectiveness of coordinated intelligence, advanced weaponry, and tactical expertise. These confrontations served as catalysts for refining and implementing innovative strategies, enabling the Allies to gain crucial technological capabilities and operational superiority.

Turning points were achieved through landmark operations that disrupted enemy supply chains, dismantled strategic infrastructure, and paralysed key research and development facilities, thereby reducing the production and proliferation of advanced technologies. The dismantling of these essential resources not only defused the enemy's technological momentum, but also reshaped the landscape of the conflict, creating conditions conducive to the unveiling and exploitation of new breakthroughs in science and engineering. These major encounters and turning points, characterised by decisive action and strategic foresight, redefined the trajectory of the Allies' pursuit, paving the way for unprecedented advances and lasting legacies in the fields of warfare and innovation.

Analysis of enemy tactics and countermeasures

In the aftermath of the Second World War, a comprehensive analysis of enemy tactics and countermeasures became imperative for the Allied forces. Understanding the intricacies of Nazi tactics and strategies employed during the war was essential to formulating overall post-war strategies and anticipating potential future conflicts. This analysis examined the methods, technologies, and ideologies that shaped the military operations of the Axis powers, providing valuable information for the development of defensive and offensive measures by the Allies. The examination of enemy tactics covered a wide range of aspects, including troop deployment, intelligence gathering, propaganda, technological innovations, and the use of unconventional warfare. By examining German army manoeuvres in various theatres of war, such as the Eastern and Western Fronts, the Pacific and Africa, the Allies gained essential knowledge about the strengths and weaknesses of their adversaries' strategies.

The study focused on the organisational structure of the Axis forces, their command and control mechanisms, and the decision-making processes that influenced their tactical deployments. At the same time, the assessment of enemy countermeasures involved dissecting the defensive responses and adaptations made by the Axis powers to counter Allied offensives. This involved examining the fortifications, anti-aircraft systems, encryption technologies, and other defensive mechanisms used by the Germans and their allies to thwart the advances of opposing forces. Through detailed evaluation of these countermeasures, the Allies were able to identify weaknesses, assess the effectiveness of their own weapons and strategies, and recalibrate their approaches for future military engagements.

The analysis of enemy tactics and countermeasures was not limited to historical documentation. It also included interviews with captured enemy personnel, inspection of confiscated equipment, and exploitation of captured territories to gather first-hand intelligence on the intricacies of Axis operations. This multidimensional approach provided insight into the evolution of enemy tactics and countermeasures throughout the conflict. The outcome of this rigorous analysis played a critical role in shaping the post-war defence and security policies of the Allied nations. It served as the basis for the development of military doctrines, the creation of intelligence agencies, and the development of technological capabilities designed to counter the potential remnants of Axis aggression and mitigate emerging threats on the international stage. The lessons learned from the insightful analysis of enemy tactics and countermeasures helped shape global geopolitical dynamics and lay the foundation for modern warfare strategies.

Impact on post-war strategies

After the tumultuous period of the Second World War, the impact on post-war strategies was profound and far-reaching. Technological advances and intelligence-gathering efforts during the war necessitated a reassessment of global power dynamics and military strategies. One of the main impacts on post-war strategies was the integration of captured enemy technologies into the arsenals and research programmes of the victorious Allied powers. The Allies recognised the critical role that technological superiority played in achieving victory. They then sought to leverage these advances to maintain a competitive advantage in the post-war period. The adoption of advanced technologies from defeated adversaries fundamentally reshaped military doctrines and defence planning around the world.

Intelligence gathered during the war, particularly regarding enemy tactics and developments, provided crucial insights that influenced post-war strategies. Meticulous analysis of enemy tactics and countermeasures allowed the Allies to learn from their adversaries' mistakes and successes, informing strategic decisions in the uncertain landscape of the post-war world. The amalgamation of this intelligence and captured enemy technologies strengthened the victorious powers' tactical and strategic repertoire, profoundly shaping their approach to future conflicts and diplomatic negotiations. The impact on post-war strategies extended beyond the immediate military sphere, as ethical considerations surrounding technological seizure and wartime innovation prompted deliberations on international norms and rules of engagement. The implications of integrating captured technologies and exploiting wartime intelligence sparked debates about the ethical limits of warfare and the responsible use of advanced weaponry and surveillance capabilities. These discussions ultimately contributed to the development of new international frameworks and agreements aimed at regulating the use of military technologies and ensuring the ethical behaviour of nations in times of conflict. Thus, the influence of the Second World War on post-war strategies is felt not only in military doctrine, but also in the ethical and legal dimensions of international relations.

The impact of post-war strategies extended beyond the immediate military context, stimulating advances in scientific research, industrial innovation and geopolitical realignments. The knowledge gained during wartime, combined with the assimilation of captured technologies, fuelled transformative developments in fields such as aerospace, computing and telecommunications. The strategic pursuit of scientific and technological supremacy became intertwined with national security imperatives, leading to unprecedented investments in research and development that continue to shape contemporary global power dynamics and economic competition.

The reorientation of post-war strategies influenced diplomatic negotiations and alliances, redefining the geopolitical landscape and setting the stage for Cold War rivalry between ideological blocs. In essence, the impact of post-war strategies transcended military considerations, exerting significant influence on post-war technological, economic and geopolitical trajectories.

In sum, the impact of the Second World War on post-war strategies highlighted the enduring importance of technological advances and intelligence gains in shaping global power dynamics and military perspectives. The assimilation of captured enemy technologies, knowledge gained from intelligence operations, and ethical considerations arising from wartime innovations collectively underpinned a paradigm shift in national defence postures and international relations. The repercussions of these strategic recalibrations permeated multiple facets of societal and geopolitical life, leaving an indelible mark on the trajectory of human history and the evolution of modern warfare.

Reflections on Long-Term Outcomes

The long-term outcomes of any historical event are inevitably complex and multifaceted, and the impact of the

Second World War on post-war strategies is no exception. When we reflect on the consequences of the war, it becomes clear that the geopolitical landscape was irrevocably altered in ways that continue to shape global politics, alliances, and international relations to this day. The lessons learned and strategies developed during the post-war period left a deep and lasting mark on the world.

One of the most notable long-term outcomes is the creation of the United Nations, an entity designed in response to the devastation caused by the war and intended to encourage cooperation, prevent conflict, and promote peace and security among nations. The principles and structures put in place after the war reflect a concerted effort to institutionalise mechanisms for collective security, diplomacy and international law, and the United Nations remains a central force in world affairs.

The post-war period saw the emergence of the Cold War, a period characterised by intense ideological rivalry and geopolitical tensions between the United States and the Soviet Union. The legacy of wartime strategies and alliances directly influenced the trajectory of this turbulent and protracted struggle, shaping military doctrines, diplomatic engagements, technological advances, and proxy conflicts around the world. The long-term repercussions of the Cold War are still felt in contemporary international politics and continue to influence the behaviour of states and non-state actors.

In addition to reshaping the geopolitical order, the post-war environment stimulated rapid development in science, technology, and innovation, particularly in the fields of aerospace, nuclear energy, and computing. The strategic imperatives of war fuelled unprecedented investment in research and development, leading to profound advances that catalysed societal transformation and economic progress. The legacy of these innovations continues to redefine the contours of modern civilisation and underpins the dynamics of global competition and cooperation. Finally, the impact of post-war strategies is also evident in the realm of decolonisation and the restructuring of colonial empires. The war precipitated changes in power dynamics, public consciousness, and anti-imperial sentiment, which ultimately facilitated the decolonisation processes that swept across Asia, Africa, and the Caribbean.

The emergence of new independent states and the reconfiguration of global power structures fundamentally recalibrated the geopolitical map, giving rise to new alliances, new frictions and new opportunities. As we consider the long-term outcomes of the post-war period, we are compelled to take into account the lasting legacies of the Second World War and the profound transformations that have indelibly shaped the modern world. The repercussions of post-war strategies continue to resonate across generations, testifying to the enduring importance of history in shaping the course of human civilisation.

8
Medicine and Morality
The Ethical Dilemma

The intersection of innovation and ethics

The historical backdrop of medical advances during wartime provides profound insight into the complex balance between innovation and ethics. Amidst the chaos and destruction of war, the quest for medical progress intensified, leading to remarkable and sometimes controversial advances. The synthesis of necessity and opportunity fuelled the rapid evolution of medical practices, pushing the boundaries of scientific knowledge and human experimentation. Thus, we are interested in the myriad complexities that emerged at the crossroads of innovation and ethical dilemmas during this tumultuous period. By examining the historical context and ethical dilemmas faced by medical practitioners, we can better understand how wartime innovations continue to shape contemporary medical ethics and principles. Understanding this dynamic intersection is essential to grasping the profound impact of historical events on today's medical practices and ethical standards.

Historical Context: Medical Advances During Wartime

In the tumultuous context of World War II, the field of medicine witnessed unprecedented challenges and remarkable advances. The demands of wartime propelled medical science into uncharted territory, ushering in an era defined by rapid innovation and ethical dilemmas. While conflict often breeds destruction, it also stimulates innovation, and the medical sphere was no exception.

In the crucible of war, the demand for effective medical care reached a critical level. Faced with unprecedented levels of casualties and injuries, physicians and surgeons were confronted with the daunting task of treating combat-related wounds, developing new surgical techniques, and advancing innovative medical technologies. War served as a catalyst for revolutionary developments in blood transfusion, triage systems, and prosthetics, as well as the introduction of antibiotics to combat previously incurable infections. The convergence of necessity and ingenuity led to prolific advances in trauma care that ultimately reshaped the landscape of modern medicine. Simultaneously, the war provided the impetus for the exploration of psychological trauma and psychiatry. The devastating toll of battle precipitated an urgent need to understand and alleviate the mental afflictions of soldiers. Thus, the war era was marked by decisive advances in the study and treatment of post-traumatic stress disorder (PTSD) and other combat-induced psychological disorders, laying the foundation for the evolution of modern psychiatric care.

War had a profound impact on public health initiatives and disease management. Epidemics, particularly in internment and refugee camps, highlighted the urgency of preventive measures and epidemiological research. The fight against infectious diseases such as malaria, typhus, and tuberculosis led to the development of vaccines, the implementation of mass sanitation campaigns, and the enactment of comprehensive public health strategies. These efforts not only preserved the health of populations amid the turmoil of war, but also left a lasting imprint on global healthcare practices.

In short, the crucible of World War II catalysed unprecedented advances in medicine, highlighting humanity's capacity for innovation in the face of adversity. This historical retrospective allows us to better understand the transformative power of war on medical science, a legacy that continues to shape healthcare ethics and ethical considerations in our contemporary society.

Human experimentation: A moral debacle

During the tumultuous period of global conflict, the field of medicine found itself mired in an ethical quagmire when unethical human experimentation took root amid the demands of wartime. The proliferation of these practices was a direct consequence of the perversion of medical advances for nefarious purposes. The moral and ethical implications of these detestable acts paved the way for profound introspection on the darkest aspects of human nature and the limits of science. Human experimentation during this period represented a flagrant violation of the fundamental principles of medical ethics, ultimately crystallising into a moral debacle that resonates throughout history. The manipulation and exploitation of vulnerable individuals for the sake of supposed improvements in warfare demonstrated a blatant disregard for the sanctity of human life. Unjustified experiments conducted on prisoners of war and innocent civilians under the guise of scientific research not only violated the Hippocratic oath, but also infringed upon the inherent rights and dignity of other human beings. This flagrant transgression against humanity highlights the dangerous precipice into which the pursuit of knowledge teetered during one of the darkest periods in modern history.

The unfathomable suffering and irreversible damage inflicted on countless individuals in these abhorrent experiments cast an indelible shadow over the ethical landscape of medical research. These transgressions serve as a poignant reminder of the catastrophic consequences that arise when ethical lapses are allowed to infiltrate the pathways of scientific inquiry. The systematic dehumanisation of subjects and disregard for their suffering underscore the monumental breach of trust between medical practitioners and the communities they were meant to serve. The repercussions of these unethical endeavours reverberate across time and space, illustrating the enduring importance of adhering to ethical principles in medical practice. As we delve into the annals of historical medical atrocities, we must confront the irrevocable damage caused by unethical pursuits and resolutely reaffirm our unwavering commitment to safeguarding the intrinsic value of every human life. It is through this introspection and collective acknowledgement of past transgressions that we strengthen the ethical foundations of modern medical research, ensuring that the shadows of malice never again darken the noble quest for healing and understanding.

Case Study Analysis: Controversial Experiments and Their Outcomes

During the tumultuous period of World War II, numerous

controversial medical experiments were conducted by various parties, often leading to serious ethical dilemmas that continue to be discussed and scrutinised today. These experiments raised crucial questions about the limits of scientific research, the rights of human subjects, and the responsibility of healthcare professionals to uphold ethical standards. One such notorious case study is the experimentation conducted at the Dachau concentration camp by the Nazi regime, where prisoners were subjected to studies on hypothermia, malaria research, and other brutal medical trials. The profound suffering inflicted on these individuals in the name of science starkly illustrates the heinous consequences of unchecked scientific ambition.

Another relevant case is that of the infamous Unit 731 of the Imperial Japanese Army, where inhumane experiments were conducted on prisoners, including vivisection, research on bacteriological warfare, and exposure to extreme conditions. The results of these experiments not only caused immeasurable harm to the victims, but also had a significant impact on medical ethics and human rights globally.

The disturbing legacy of these case studies has prompted profound reflection on the need for rigorous ethical oversight and the imperative to protect the fundamental dignity of all human beings. Detailed analysis of each case study reveals the disconcerting juxtaposition of relentless scientific pursuit and the intrinsic value of human life. By examining the ethical implications and ramifications of these experiments, we gain valuable insights into the complexity of balancing scientific progress with moral obligations. These case studies serve as a cautionary tale, highlighting the critical need for strict ethical guidelines, robust oversight mechanisms, and an unwavering commitment to safeguarding the well-being of those involved in medical research. As we navigate the complex terrain of bioethics and research ethics, a thorough exploration of these contentious experiments and their consequences allows us to better understand the profound ethical dilemmas inherent in the pursuit of medical progress.

The Legal Framework: The Nuremberg Code and Beyond

The aftermath of World War II marked a turning point in medical ethics with the introduction of the Nuremberg Code. This historic document, formulated in response to the heinous human experiments conducted by Nazi physicians during the war, established the fundamental principles of ethical medical research involving human subjects. The ten points of the Nuremberg Code include essential requirements such as the voluntary consent of the subject, the absence of unnecessary physical or mental suffering, and the need for the research to produce fruitful results for the good of society. This fundamental framework emphasises, in particular, the imperative to prioritise the well-being and rights of individual participants in any form of medical research. By emphasising informed consent and the need to minimise risks, the Nuremberg Code ushered in a new era of ethical considerations in medical practice and research. Building on the Nuremberg Code, subsequent developments in international and national regulations have strengthened the legal landscape surrounding medical ethics. The Declaration of Helsinki, first adopted by the World Medical Association in

1964, introduced guidelines for regulating non-therapeutic research and established standards for clinical trials. In the United States, the Belmont Report consolidated the principles of respect for persons, beneficence, and justice as the cornerstones of ethical research. The creation of institutional review boards (IRBs), responsible for safeguarding the rights and welfare of human subjects, marked a crucial step in the implementation of ethical standards at the institutional level.

Several pieces of legislation, such as the US Common Rule and the EU Data Protection Directive, have contributed to the evolution of global regulations on research involving human subjects. Today, advances in areas such as genomic research and artificial intelligence have sparked ongoing discussions and revisions of ethical guidelines to ensure the protection of participants and the integrity of scientific research. As the frontiers of medical research continue to expand, the ethical ramifications require ongoing review and adaptation of legal frameworks to maintain the highest standards of patient welfare and human dignity.

Bioethics: Lessons learned from the atrocities of the past

The atrocities committed during the Second World War prompted significant reflection on medical ethics and the need for strict regulations to prevent further transgressions. The Nuremberg Code, established in 1947 following the trials of Nazi doctors, defined the fundamental principles governing human experimentation and laid the foundation for contemporary bioethical frameworks. The Nuremberg Code emphasised the need for voluntary consent, the need to avoid unnecessary suffering, and the importance of scientific validity in research, thus setting a precedent in medical ethics.

However, it became clear that the Nuremberg Code had limitations, as it focused primarily on regulating research on humans, often neglecting broader ethical considerations in medicine. Subsequent events, such as the revelations of the Tuskegee syphilis study and the Willowbrook experiments, highlighted the need to continually refine bioethical guidelines. These incidents underscored the ethical obligation to respect patient autonomy, maximise benefits, and minimise harm in all facets of medical practice. The historical origins of unethical medical practices have highlighted the profound influence of power dynamics, discrimination, and exploitation in clinical research and healthcare delivery. The pursuit of comprehensive bioethical standards has required recognition of these societal inequalities and an unwavering commitment to justice and fairness in medical decision-making. As a result, contemporary bioethics has evolved beyond the confines of research ethics to encompass broader concerns such as distributive justice, cultural competence, and equity in healthcare.

The landmark publication of the Belmont Report in 1979 reinforced the ethical precepts guiding research involving human subjects, emphasising respect for persons, beneficence, and justice. The report's ethical principles continue to reverberate across various medical disciplines, reinforcing the enduring importance of ethical integrity in all dimensions of healthcare. In addition, the establishment of institutional review boards and ethics oversight committees has played a critical role in ensuring compliance with ethical standards and fostering a culture of accountability in medical institutions. The trajectory of bioethics is characterised by a constant re-evaluation of moral imperatives and ethical dilemmas, propelling the current discourse on medical ethics into contemporary landscapes. By critically evaluating historical errors and strengthening ethical safeguards, society can consolidate its commitment to ethical probity, fostering a healthcare environment based on compassion, integrity, and respect for human dignity.

Post-war medical practices and protocols

After the end of the Second World War, the medical community was faced with the daunting task of reconciling the ethical lapses and human rights violations that occurred during the war. The atrocities committed under the guise of scientific research highlighted the need for rigorous medical practices and protocols. As the world grappled with the consequences of wartime experimentation, a concerted effort was made to reform medical ethics and ensure that such injustices would never happen again. In the post-war period, the importance of informed consent and patient autonomy underwent a notable evolution. Medical professionals and researchers recognised the fundamental right of individuals to make decisions about their own bodies and treatment options. This fundamental change was a decisive step in defending the dignity and rights of patients in the medical field.

The establishment of ethics review boards and institutional oversight mechanisms became an integral part of medical practice. These entities were tasked with evaluating proposed research studies to ensure compliance with ethical standards and the protection of human subjects. By subjecting research protocols to rigorous review, these measures aimed to prevent the recurrence of unethical practices that had tarnished the field of medicine. Post-war medical practices also underwent a significant evolution with the introduction of regulations regarding confidentiality and privacy. Protecting patient information and maintaining the confidentiality of medical records became paramount, fostering trust and accountability within the healthcare system. The enforcement of strict privacy standards also helped to respect individual rights to privacy and strengthen public confidence in the medical profession.

The field of medical education has been transformed, incorporating ethics education as an essential component of training programmes. By instilling ethical principles and moral reasoning into the curriculum, future healthcare professionals have been equipped to deal with complex ethical dilemmas and uphold the highest standards of integrity in their professional activities. As the legacy of wartime injustices reverberated throughout the medical community, the post-war period heralded a paradigm shift towards prioritising ethical considerations in all facets of medical practice. This profound evolution laid the groundwork for the development of comprehensive ethical frameworks that continue to shape contemporary medical practices and protocols.

Evolution of patient rights and informed consent

Patient rights and the concept of informed consent underwent significant evolution in the aftermath of World War II, particularly in response to the ethical dilemmas raised by wartime medical practices and human experimentation. The recognition of individual autonomy and the right to make informed decisions about one's own medical care became a cornerstone of modern medical ethics, coinciding with an increased emphasis on transparency and accountability within the healthcare system. The post-war period saw the emergence of landmark declarations and conventions, such as the Nuremberg Code and the subsequent development of the Declaration of Helsinki, which set out ethical principles governing experimentation and research on human subjects. These fundamental documents laid the groundwork for the establishment of ethical standards regarding patient rights and informed consent, emphasising the need for voluntary and well-informed consent from individuals participating in medical research or clinical trials. As medicine advanced, so too did the understanding of patient rights and the complexities surrounding informed consent.

One of the key moments in the evolution of patient rights was the recognition that informed consent should not only encompass the provision of information to patients, but also their understanding of that information. This shift towards ensuring genuine understanding marked a crucial step in safeguarding patient autonomy and well-being, reflecting a more nuanced approach to the concept of informed consent.

Ongoing ethical discourse and debate have refined the

processes of informed consent, taking into account cultural diversity, language barriers, and varying levels of health literacy among patient populations. Recognising the multifaceted nature of informed decision-making, healthcare professionals and researchers have sought to develop strategies that account for these complexities while respecting the fundamental principles of patient autonomy and beneficence.

Today, the landscape of patient rights and informed consent continues to evolve in response to emerging issues and advances in medical technology. The proliferation of personalised medicine, genomics, and innovative treatments has introduced new considerations for informed consent, raising questions about the implications of genetic information, the use of experimental therapies, and the potential intersection of individual and collective interests. These developments underscore the ongoing importance of critically examining and adapting ethical frameworks to ensure that patient rights and informed consent remain robust and relevant in the evolving healthcare landscape. In essence, the evolution of patient rights and informed consent reflects an enduring commitment to respecting the dignity and autonomy of individuals in the field of medicine. As we navigate the complexities of modern healthcare, the ongoing re-examination and refinement of the principles surrounding patient rights and informed consent are essential to preserving ethical integrity and promoting the well-being of individuals receiving medical care.

Current Ethical Challenges in Medicine

As we embark on a journey into the realm of modern medicine, it is imperative to address the ever-evolving ethical challenges that continue to shape the healthcare landscape. The intersection of technological advances, patient rights, and the pursuit of medical innovation has given rise to complex ethical dilemmas that require careful navigation. One of the key challenges lies in the ethical implications of genetic engineering and gene editing. With the development of CRISPR technology, the possibility of modifying the human genome raises profound ethical concerns about the limits of scientific intervention and the impact on future generations.

The rapidly growing field of personalised medicine introduces ethical considerations related to privacy, data security, and the equitable distribution of resources. The possibility of tailoring medical treatments to an individual's genetic makeup poses unprecedented ethical challenges regarding informed consent, cost-effectiveness, and accessibility. In addition, the growing importance of artificial intelligence (AI) in healthcare raises ethical questions about autonomy, accountability, and potential biases inherent in algorithmic decision-making. While AI promises to revolutionise diagnostics and treatment protocols, ethical frameworks must be established to ensure transparency, fairness, and patient well-being.

The ethical conundrums surrounding end-of-life care and medically assisted suicide remain controversial issues at the intersection of medicine and morality. The debate over patient autonomy, quality of life, and the role of healthcare professionals in these decisions continues to spark ethical introspection and advocacy for legislative clarity. Beyond these specific challenges, the broader dynamics of healthcare delivery, resource allocation, and disparities in access to care underscore the pervasiveness of ethical considerations in the fabric of modern medicine. Addressing these challenges requires interdisciplinary collaboration, ongoing ethical discourse, and a firm commitment to the principles of beneficence, non-maleficence, justice, and respect for autonomy. As medical technologies and societal values evolve, a proactive engagement with these ethical challenges is paramount to ensuring that advances in medicine align with the overarching goal of improving human well-being.

Conclusion: Future Directions in Medical Ethics

As we stand at the crossroads of medical ethics, it is essential to consider the future trajectory of ethical standards and principles in the field of medicine. The dynamic landscape of healthcare, with its rapid technological advances and evolving societal norms, presents a myriad of challenges and opportunities for ethical considerations. To shape the future of medical ethics, several key directions deserve special attention. One relevant aspect concerns the integration of emerging technologies, such as artificial intelligence and genetic engineering, into medical practices. As these innovations continue to redefine the frontiers of healthcare, it is imperative to establish robust ethical frameworks that govern their implementation, ensuring beneficence, non-maleficence, autonomy, and justice.

The increasing globalisation of healthcare requires a comprehensive approach to resolving ethical dilemmas arising from cultural variations, resource disparities, and differing legal regulations across countries. Global collaborative efforts are essential to establish universally applicable ethical guidelines while respecting diverse cultural sensitivities and values.

The rapidly expanding field of precision medicine requires a reassessment of ethical standards regarding informed consent, data confidentiality, and equitable distribution of cutting-edge treatments. The ethical implications of gene editing, personalised therapies, and predictive diagnostics underscore the urgency of proactive ethical governance to safeguard patient rights and well-being. Furthermore, as we navigate an era of unprecedented medical discoveries and interventions, the ethical responsibilities of healthcare professionals, researchers, and industry stakeholders are becoming increasingly complex. Striking a balance between innovation and ethical practice requires a firm commitment to transparency, integrity, and ethical leadership at all levels of the healthcare ecosystem. It is imperative to foster a culture of ethical awareness and continuing ethics education within medical institutions to create the foundation for conscientious decision-making and professional conduct.

Finally, the ongoing pursuit of justice and equity in healthcare requires constant re-examination of systemic inequalities, vulnerable populations, and access to essential medical resources. By addressing the social determinants of health, advocating for universal access to healthcare, and confronting institutional biases, the ethical compass of medicine aligns with the core principles of equity, compassion, and social responsibility. In summary, the future of medical ethics depends on proactively adapting to technological advances, cultural diversity, the challenges of precision medicine, ethical professionalism, and equitable healthcare paradigms. By embracing this ever-changing landscape with foresight and moral integrity, we can navigate the complex ethical terrain of healthcare and cultivate a future where ethical imperatives underpin all facets of medical practice and innovation.

9
The Secrets Of Special Operations

Overview of Special Operations during the Second World War

The Second World War marked a decisive stage in the evolution of special operations, spearheading the development of clandestine warfare on a global scale. As nations turn to unconventional tactics to gain strategic advantages, the genesis of special operations units emerges from the crucible of intense conflict and the changing dynamics of modern warfare. The origin of these specialised units can be traced back to the urgent need for intelligence gathering, sabotage and subversion behind enemy lines, which required a radically different approach to conventional military strategies. To meet this demand, several nations created secret organisations and elite forces, leveraging the expertise of extraordinary individuals and innovators who operated with unparalleled discretion and precision. The creation of these units was intended to disrupt enemy supply lines, deceive military forces, and gather crucial information to influence the outcome of the war. The global context of Second World War special operations units highlights a complex tapestry of geopolitical tensions, espionage, and daring exploits that forever changed the landscape of modern warfare. Here, we focus on the genesis and evolution of these remarkable units, highlighting the strategic foresight and daring execution that defined clandestine operations during the Second World War.

Strategic Planning and Execution

Strategic planning and execution of special operations during the Second World War were essential elements of the Allies' efforts to gain an advantage over their adversaries. Special operations involved meticulous planning, coordination, and precise execution, often in clandestine and high-risk environments. They required a thorough understanding of the enemy's strengths and weaknesses, as well as the ability to adapt to rapidly changing situations. The success of these operations depended on the strategic acumen of military leaders and the bravery of agents who risked their lives behind enemy lines.

A key aspect of strategic planning was the identification and selection of targets that could provide a significant tactical advantage. Special operations units meticulously gathered intelligence, often through daring reconnaissance missions, to identify high-value targets such as crucial supply routes, communication centres, or key enemy personnel. Once these targets were identified, meticulous planning was put in place to develop strategies for infiltrating, sabotaging, or neutralising them. This required a thorough understanding of enemy tactics and defences, as well as innovative approaches to overcoming them. The execution of special operations demanded precision and discretion. Operators underwent rigorous training to acquire specialised skills in infiltration, sabotage and espionage. Each operation was rehearsed meticulously, with particular attention paid to contingency plans and unforeseen challenges.

Effective communication and coordination between allied

forces were essential to the flawless execution of these operations. The success of the operations depended on the discretion of the agents, their ability to adapt to changing circumstances, and the support network that facilitated their missions. In addition, strategic planning included the use of diversionary tactics to deceive the enemy and create opportunities for successful operations. Deception and misdirection are integral to many special operations, allowing agents to exploit the element of surprise and confuse the enemy. Whether through disinformation campaigns or orchestrated diversions, strategic planners sought to destabilise the enemy and maximise the chances of mission success. The importance of strategic planning and execution in special operations extends beyond the battlefield. It involves orchestrating multidisciplinary expertise, including intelligence gathering, logistics, technology, and human psychology. It is essentially the fusion of precise strategy and bold execution to achieve decisive results. The lessons learned from these operations continue to influence modern military tactics and strategic planning, demonstrating the lasting impact of innovation and ingenuity in wartime.

Key figures in covert operations

During the Second World War, the success of covert operations relied heavily on the expertise and leadership of key figures in the field of intelligence and espionage. These individuals played a vital role in the outcome of many clandestine missions, often working behind enemy lines to gather crucial information and orchestrate strategic strikes. Among the most notable figures was the charismatic and enigmatic Virginia Hall, an American spy who overcame great adversity to become one of the most effective agents of the war. Known for her exceptional cunning and ingenuity, Virginia Hall operated in Nazi-occupied France, where she coordinated vital resistance efforts and provided invaluable intelligence to the Allies. Her courage and unwavering commitment to the cause made her a legend in the world of covert operations. Another influential figure was British intelligence officer Major Airey Neave, who played an important role in developing escape and evasion tactics, as well as coordinating anti-German resistance activities. Neave's bold and innovative approach helped shape covert operations and set new standards for intelligence gathering and subversive activities.

The brilliant mind of William Stephenson, also known as *Intrepid*, stands out as a central figure in the field of special operations. As head of British security coordination, Stephenson led numerous covert and intelligence operations throughout the war, laying the groundwork for modern espionage practices. His strategic vision and unparalleled contributions earned him the admiration of his allies and struck fear into the hearts of his adversaries. These key figures, along with many others, contributed to the success of covert operations during a time of unprecedented global upheaval. Their unwavering dedication, astute leadership, and courageous actions continue to demonstrate the lasting impact of their contributions.

Technological Tools and Innovations

During the Second World War, technological advances played a vital role in the evolution of warfare. Innovation and the application of new technologies greatly influenced the outcomes of covert operations and espionage activities. As such, we pay particular attention to the technological tools and innovations that revolutionised clandestine missions during this tumultuous period. One of the most significant technological advances was the development of miniaturised cameras and recording devices, which enabled agents to discreetly gather valuable intelligence without compromising their cover. These compact but powerful tools paved the way for unprecedented surveillance capabilities, allowing secret agents to gather crucial information while operating in hostile environments.

Advances in cryptography and encryption techniques provided a crucial advantage in securing sensitive communications and protecting classified data from enemy interception. The complex design of encryption machines and devices enabled the secure transmission of vital messages, preserving the integrity of clandestine operations.

The integration of radio technology into espionage activities revolutionised communication and coordination between secret agents. Portable and concealable radio transmitters facilitate the seamless transmission of essential information, allowing agents to stay in contact with their superiors and relay vital information from the field. The development of stealth and concealment technologies has also played a critical role in improving the effectiveness of clandestine missions. Advanced camouflage materials and disguises have enabled agents to blend into their surroundings, evade detection, and increase the success rate of covert operations.

Advances in clandestine transportation, such as miniature submarines and specialised aircraft, provided clandestine agents with unprecedented mobility and infiltration capabilities. These innovative transportation solutions allowed agents to access remote and fortified locations, expanding the scope and impact of their missions. The convergence of technological ingenuity and strategic innovation increased the effectiveness and impact of clandestine operations during World War II, laying the foundation for modern espionage practices. The legacy of these technological advances continues to influence the evolution of the clandestine mission landscape, demonstrating the enduring importance of innovation in the field of espionage and special operations.

Psychological Warfare Tactics

During the Second World War, psychological warfare tactics took on considerable importance as a means of gaining strategic advantage and influencing the behaviour of enemies and allies. These tactics sought to exploit the psychological vulnerabilities and fears of adversaries, using a range of methods to sow confusion, disunity and demoralisation. One of the best-known tactics was the dissemination of propaganda through leaflets, radio broadcasts and other media channels, designed to undermine the morale of enemy forces and civilians while bolstering friendly populations.

Covert operations use psychological tactics, such as spreading false rumours and misinformation, to mislead and disorient the enemy, creating uncertainty and disrupting their decision-making process. The use of deceptive imagery, including camouflage and decoy operations, also played a key role in psychologically manipulating adversaries' perceptions. Special units were trained in psychological operations, using various forms of persuasion and manipulation to achieve military objectives without direct combat. These units used fear, uncertainty, and doubt to weaken the resolve of opposing forces, often relying on cultural and social nuances to reinforce the impact of their messages.

The psychological impact of unconventional warfare, such as guerrilla tactics and subversive activities, cannot be underestimated. By striking at enemy morale and instilling doubt about the effectiveness of their leaders, psychological warfare tactics contributed significantly to the overall success of military campaigns. The enduring legacy of these tactics is evident in modern psychological warfare strategies, which continue to be employed by military and intelligence agencies to shape perceptions, influence behaviour, and achieve strategic objectives in contemporary conflicts and geopolitical engagements.

Case Studies: Successful Missions

Throughout the turmoil of World War II, various clandestine missions were carried out with incredible precision and unwavering commitment, ultimately shaping the trajectory of the war. These specialised operations represented the pinnacle of strategic warfare, where meticulous planning and covert execution played a vital role in changing the course of history. The successful missions orchestrated during this period are a testament to the extraordinary bravery, intelligence, and determination of those involved. One such remarkable mission was Operation Grouse, a daring Norwegian operation that laid the groundwork for the subsequent sabotage of the Nazi heavy water production facility at Vemork.

The daring exploits of the Norwegian commandos exemplified the meticulous coordination and flawless execution required for such undertakings. Despite difficult terrain and overwhelming adversity, they displayed unparalleled resilience and bravery, succeeding in disrupting a key aspect of the German atomic bomb project. The raid on Saint-Nazaire, often considered one of the most daring and reckless naval operations of the war, is another compelling example. British commandos executed a meticulously planned and highly risky assault on the heavily fortified dry dock, strategically targeting the essential infrastructure that supported Germany's formidable battleships. The commandos' unfathomable bravery, combined with decisive tactical prowess, resulted in a resounding success that significantly hampered Nazi naval capabilities.

The clandestine efforts of the Office of Strategic Services (OSS) also merit attention, particularly the exceptional achievements of the Jedburgh teams. Composed of elite American, British, and French soldiers, these teams undertook perilous missions deep behind enemy lines, leveraging their expertise in unconventional warfare to support resistance movements and disrupt German operations. Their clandestine successes reverberated throughout occupied

Europe, instilling hope and courage in populations besieged by oppression and tyranny. These exemplary missions underscore the indomitable spirit of those dedicated to preserving freedom and catalyse crucial shifts in the tides of war. They stand as a lasting testament to the invaluable impact of covert operations on the grand stage of history, illustrating the profound influence of strategic finesse, unwavering determination, and consummate bravery.

Communication Systems and Secrecy

During World War II, one of the key elements that ensured the success of special operations was the robustness of communication systems and the absolute secrecy maintained by undercover agents. Communication is central to military operations and intelligence activities, as it has a direct impact on the success or failure of missions. The need for secure, reliable and instantaneous communication channels led to the development of advanced encryption technologies and methods. Special operations units rely heavily on sophisticated encoding and decoding mechanisms to transmit sensitive information without interception. These systems have played a vital role in maintaining the security and confidentiality of operations, preventing vital intelligence from falling into enemy hands. The level of secrecy surrounding these communication processes was unparalleled, with agents meticulously following protocols to avoid compromising sensitive missions. Encryption techniques such as the Enigma machine and the Lorenz cipher enabled clandestine communications, forming the backbone of covert operations. In addition to cryptographic solutions, special operations also utilised innovative signalling methods, including Morse code and radio transmissions, to relay critical messages securely.

Multiple levels of authentication and identity verification were employed to ensure that only authorised personnel could access classified communications. The complexity of these communication systems and the rigorous measures taken to protect classified information illustrate the unprecedented commitment to operational security observed during the Second World War. The ability to effectively conceal and transmit sensitive data contributed significantly to the success of clandestine missions, reinforcing the importance of communication and secrecy in specialised wartime operations.

Challenges and Setbacks

Following clandestine operations during World War II, it became clear that special operations were not immune to difficulties and setbacks. Despite meticulous planning and execution, unforeseen obstacles often arise, testing the resilience and adaptability of the agents involved. One of the main challenges was the constant risk of compromise and infiltration by enemy agents. The complex network of communication systems and secrecy measures, while largely effective, was always susceptible to flaws. The constant threat of adversaries intercepting or deciphering sensitive messages posed a formidable challenge to maintaining operational security.

The extensive use of covert methods meant that any failure could have disastrous consequences and lead to catastrophic results, underscoring the immense pressure faced by those tasked with executing these missions. Another significant setback stems from the unpredictable nature of wartime conditions, where external factors such as weather, geopolitical changes, or unexpected enemy movements can disrupt carefully laid plans.

The ethical dilemmas inherent in special operations often posed complex problems. Balancing the need to achieve strategic objectives, moral considerations, and potential collateral damage required an unwavering commitment to principles amid the turmoil of conflict. Furthermore, the psychological impact on agents participating in covert missions should not be overlooked. The constant tension of operating under the radar, often in hostile territories and highly stressful environments, took a toll on the mental well-being of those serving their respective nations. In light of these challenges and setbacks, the resilience and ingenuity demonstrated by those involved in special operations at that time are a testament to the strength of character and dedication of the human spirit, despite the arduous and discouraging circumstances they faced.

Impact on post-war espionage techniques

The impact of the Second World War on post-war espionage techniques cannot be overstated. The war served as a breeding ground for the development and refinement of covert operations, intelligence gathering and surveillance tactics, which profoundly influenced the global espionage landscape in the decades that followed. Lessons learned from clandestine activities during the war directly shaped the strategies and methods employed by intelligence agencies and secret agents during the Cold War and beyond. One of the most significant impacts of the Second World War on post-war espionage techniques was the increased emphasis placed on technological advances. The conflict led to rapid innovations in communication systems, encryption methods, and surveillance technologies. These developments not only revolutionised the way intelligence was gathered, but also enabled the creation of advanced and sophisticated espionage tools and equipment, such as hidden recording devices, secret cameras and code-breaking machines. The integration of cutting-edge technologies fundamentally transformed the operational capabilities of intelligence agencies, ushering in an era of unprecedented sophistication in espionage practices.

The experience of war prompted a reassessment of the role of human intelligence (HUMINT) in espionage operations. Successful infiltration and sabotage missions carried out by special operations forces during World War II highlighted the critical importance of recruiting and training skilled agents to gather intelligence behind enemy lines. This shift in focus towards clandestine human intelligence laid the groundwork for the expansion of covert agent networks and the development of specialised training programmes designed to cultivate the next generation of undercover agents.

Ethical considerations arising from the conduct of espionage during the war had a lasting influence on post-war intelligence operations. Awareness of the moral dilemmas faced by intelligence agents involved in clandestine activities led to a renewed emphasis on ethical guidelines and codes of conduct within intelligence services. The need to balance national security imperatives with ethical standards became a key concern that shaped the evolution of post-war espionage practices and decision-making processes.

In short, the impact of the Second World War on post-war espionage techniques was considerable, multifaceted and lasting. The convergence of technological advances, strategic ideas and ethical considerations arising from wartime espionage profoundly shaped the trajectory of intelligence operations in the post-war period. The legacies of innovation, adaptation, and ethical reflection continue to resonate in the contemporary landscape of global espionage, testifying to the indelible mark that the Second World War left on the evolution of clandestine intelligence practices.

Ethical considerations and controversies

The post-war period brought to light a myriad of ethical considerations and controversies arising from the espionage techniques developed during the Second World War. As the world grappled with the aftermath of the conflict, nations found themselves confronted with the moral implications of the espionage activities that had helped shape the outcome of the war. The use of covert operations, including surveillance, infiltration, and sabotage, raised profound ethical questions about the balance between national security imperatives and individual freedoms. The use of deceptive practices and psychological warfare tactics raised concerns about the manipulation of public opinion and the potential erosion of trust within societies. One of the main controversies has centred on the impact of espionage on international relations and diplomatic trust. The revelation of clandestine intelligence-gathering operations has strained relations between former allies and fuelled suspicion between nations, leading to a paradigm shift in global diplomacy. The ethical implications of this change have been profound, as it has forced policymakers to re-evaluate traditional standards of governance and the limits of acceptable conduct in the pursuit of national interests.

The emergence of new technologies and communication systems in the field of espionage sparked debates about privacy rights and the limits of government surveillance, laying the groundwork for future ethical dilemmas concerning security and individual autonomy. Another key aspect of post-war ethical considerations was the treatment of captured agents and the use of interrogation methods. The moral dimensions of these practices came to the fore as the international community sought to establish protocols for the humane treatment of prisoners and detainees. The revelation of covert paramilitary actions and targeted assassinations also prompted scrutiny of the ethical constraints governing the conduct of intelligence agencies and military special operations units. These revelations sparked public debate about the accountability and oversight mechanisms needed to prevent abuses of power and ensure compliance with ethical standards.

The legacy of wartime espionage activities has intersected with broader societal debates about truth, transparency, and propaganda. The deliberate dissemination of misinformation and the cultivation of false narratives have highlighted the inherent tension between the imperative of national security and the preservation of democratic values. The spectre of state-sponsored disinformation campaigns and subversion tactics has cast a long shadow over public discourse, prompting calls for greater transparency and ethical clarity in the conduct of intelligence operations.

In short, the ethical considerations and controversies surrounding post-war espionage techniques have highlighted the complex interplay between national security imperatives, individual rights, and international norms. The lasting impact of these ethical dilemmas continues to shape discourse on espionage and covert operations in the contemporary era, serving as a reminder of the complex ethical terrain navigated by those involved in the shadowy world of intelligence and special operations.

10
The Enigma of Cryptography
Codes and Secrets

Cryptography during the Second World War

Cryptography played a central role in the Second World War, influencing the strategies and outcomes of the conflict on a global scale. From the outset of the war, both the Axis and Allied powers recognised the imperative of protecting their communications from interception and decryption by enemy forces. The use of codes and ciphers became both a means of transmitting confidential information and a form of intelligence warfare. The ability to intercept, decipher, and exploit encrypted messages could alter the course of battles and campaigns. Cryptography thus emerged as an essential tool for gaining strategic advantage and maintaining operational secrecy. The impact of cryptography extended beyond simple communications security. It influenced decision-making at the highest levels of command, often determining the timing and coordination of military operations. The Enigma machine, used by the German armed forces to encrypt sensitive messages, is a testament to the complexity and sophistication of cryptographic technology at that time. Its evolution and the subsequent efforts to decipher its codes illustrate the fierce intellectual battle that characterised cryptanalysis during the Second World War.

The effectiveness of cryptographic systems in protecting sensitive information profoundly influenced the development of tactical and strategic warfare. Both sides sought technological superiority in the encryption and decryption of messages, leading to the constant improvement and diversification of cryptographic systems. Understanding the strengths and weaknesses of the adversary's cryptographic methods became essential to formulating effective countermeasures and exploiting enemy communications. This contest of intelligence and innovation significantly influenced the progression and resolution of crucial military engagements. In essence, cryptography during the Second World War served as a pivot point in shaping the dynamics of modern warfare. It highlighted the importance of information security as a force multiplier, exerting considerable influence on the conduct of military operations and the outcome of the conflict. The quest for advantage through cryptanalysis advanced technology and intelligence gathering, leaving an indelible mark on the annals of history.

Evolution of the Enigma Machine

The evolution of the Enigma machine is a testament to the ingenuity and complexity of German cryptographic technology during the Second World War. Developed by German engineer Arthur Scherbius in the early 20th century, the Enigma machine was initially used for commercial purposes, offering a level of security sought after by banks and businesses. However, its use for military communications during the war made it a symbol of secrecy and strength, posing a formidable challenge to Allied code breakers. The Enigma machine underwent several iterations, with each version introducing new features and complexities to thwart decryption efforts. These advances included additional rotors, plugboard configurations, and changing wiring patterns, making Enigma increasingly complex and elusive.

Efforts to decipher the Enigma code faced great difficulties, as the machine offered 150,000,000,000,000,000,000,000 possible settings, making brute force attacks unfeasible. The German military also implemented strict operational procedures, such as daily key changes and specific message formats, adding layers of complexity that further thwarted Allied attempts at interception and decryption.

The introduction of the fourth rotor was a turning point in the evolution of Enigma, increasing the potential combinations and intensifying the cryptographic challenge for Allied cryptanalysts. Despite these obstacles, remarkable breakthroughs were achieved through the collaboration of brilliant minds, particularly at Bletchley Park, where the decryption skills of Alan Turing and others played a decisive role in cracking the secrets of Enigma. The evolution of Enigma not only symbolised technical sophistication, it also highlighted the constant struggle between encryption and decryption capabilities. As the Allies adapted their tactics and technologies, Enigma continued to evolve in an ongoing cryptographic arms race, illustrating the ever-changing landscape of wartime intelligence and technological innovation. Ultimately, the saga of the Enigma machine is a fascinating tale of human perseverance, intelligence, and the perpetual quest for knowledge amid the fog of war.

Allied Forces: Decoding Tactics

During the Second World War, the Allied forces faced a daunting challenge: deciphering the complex codes and encryption systems used by the Axis powers to secure their communications networks. This task required unwavering determination, innovative thinking, and collaboration between cryptanalysis and intelligence experts. During the war, Allied cryptanalysts used a variety of tactics to decipher enemy codes, which had a significant impact on the outcome of key battles and strategic decisions. One of the main tactics used by the Allied forces was to create decryption teams composed of mathematicians, linguists, and logisticians who worked tirelessly to decipher intercepted enemy messages. These teams meticulously analysed the patterns and frequencies of coded transmissions, often exploiting known weaknesses in the encryption methods used by the Axis powers. Through statistical analysis, frequency counting, and pattern recognition, Allied cryptanalysts were able to obtain valuable information that led to breakthroughs in decoding enemy communications.

The use of captured cryptographic equipment, such as code books and machines, provided crucial intelligence for the Allies' decryption efforts. Strategic operations were therefore necessary to acquire these materials, often through daring missions and espionage activities behind enemy lines. The information obtained from these documents enabled cryptanalysts to study and reverse-engineer cryptographic devices, revealing vital secrets and providing a better understanding of enemy encryption techniques.

Collaboration between Allied intelligence services and resistance movements in occupied territories played a vital role in obtaining valuable cryptographic information. These clandestine partnerships led to the interception of codes and equipment, providing essential clues that helped advance decryption efforts. The information gathered through these channels not only helped to decipher enemy communications, but also disrupted Axis operations and protected Allied forces. As the war raged on, the development and refinement of specialised cryptanalysis tools and techniques became essential to the Allied forces. Advanced computing machines, such as the British Bombe and the American SIGABA, automated certain decryption tasks, greatly speeding up the decryption process.

The use of probabilistic and mathematical models enabled cryptanalysts to devise more effective strategies for tackling the increasingly sophisticated cryptographic systems put in place by the Axis powers. The successful decoding of enemy messages by Allied forces had a direct influence on military campaigns and strategic decisions. From intercepting crucial orders and battle plans to identifying enemy supply routes and troop movements, the decrypted intelligence provided by Allied cryptanalysts had a profound impact on the course of the war. Ultimately, the tireless dedication, strategic ingenuity, and unwavering perseverance of the Allied decryption efforts played a vital role in securing victory and shaping the post-war landscape.

Secrets Beyond Enigma: Other Cryptographic Devices

While the Enigma machine is widely recognised for its importance in World War II cryptography, it is essential to acknowledge that other lesser-known cryptographic devices were used during this period. These devices, though overshadowed by the fame of the Enigma machine, played a vital role in maintaining the secrecy and confidentiality of various military and intelligence communications. Among these lesser-known cryptographic devices was the Japanese PURPLE machine, also known as the Type B cipher machine. Developed by Japan in the 1930s, the PURPLE machine used a unique rotor-based system, similar to Enigma, and was successful in encrypting sensitive diplomatic messages. Its complexity posed a formidable challenge to Allied code breakers. The German Lorenz SZ40/42 machine, known as Tunny by British code breakers, was a sophisticated teletype encryption machine used by the German high command. Its complexity rivalled that of Enigma and posed a significant challenge to cryptanalysis.

Hagelin cipher machines, primarily the C-36 and M-209, were widely used by American and Allied forces for field encryption. The portable and robust nature of Hagelin machines made them indispensable for securing tactical communications on the battlefield. These machines not only demonstrated the diversity of cryptographic techniques employed during the war, but also highlighted the ongoing arms race between adversaries in decryption and encryption technologies. Understanding the existence and impact of these other cryptographic devices allows us to better appreciate the multidimensionality and complexity of wartime cryptology. Beyond the shadow of the Enigma machine, these devices contributed significantly to the evolution of the history of cryptography during the Second World War and its lasting legacy in the field of information security.

Cryptanalysis: The Science and Art

Cryptanalysis, a term derived from the Greek words 'krypto' for 'hidden' and "analysis" for 'examination,' is the complex science and art of deciphering secret codes and ciphers. In the context of the Second World War, cryptanalysis played a crucial role in unravelling the clandestine communications of enemy forces, providing invaluable intelligence to the Allied powers. This complex discipline involves a combination of mathematical prowess, linguistic acuity, and a keen understanding of patterns and probabilities. Cryptanalysts meticulously examined intercepted ciphers, seeking to identify weaknesses or exploitable elements in the cryptographic systems used by adversaries. The arduous task of cryptanalysis demanded an extraordinary level of intellectual rigour, perseverance, and sheer determination. It required brilliant minds capable of unravelling the complexities of encrypted messages, often under tight deadlines and relentless wartime pressure.

Cryptanalysis was not only a reactive practice, but also a proactive endeavour, involving the development of innovative techniques and technologies to gain an advantage in the intelligence war. Successes in cryptanalysis led to decisive breakthroughs, giving the Allied forces invaluable strategic advantages. Deciphering enemy codes and encryption algorithms revealed crucial information about troop movements, naval operations, and clandestine plans, significantly influencing the course of major military engagements. Thanks to the dedicated efforts of cryptanalysts, the fog of secrecy surrounding enemy communications was pierced, providing deep insight into the enemy's intentions and capabilities. The enigmatic veil of cryptographic secrecy was lifted through the meticulous efforts of these unsung heroes, whose contributions remain etched in the annals of history. Beyond its importance in wartime, the legacy of cryptanalysis extends to contemporary cryptography, laying the foundation for advances in encryption methods and cybersecurity. The lessons learned from cryptanalysis in the past continue to reverberate in the modern era, shaping the evolution of secure communication and information protection. The complex science and art of cryptanalysis is a testament to the indomitable human spirit, the relentless pursuit of knowledge, and the enduring impact of intellectual fortitude.

Role of Cryptography in Intelligence and Espionage

Cryptography played a vital role in intelligence and espionage during and after the Second World War, influencing the outcomes of crucial missions and shaping the course of history. In the clandestine world of espionage, cryptography emerged as an essential tool for encoding sensitive information, communications, and strategic plans. The use of cryptographic techniques enabled intelligence agencies to protect their transmissions from interception by hostile forces, ensuring the confidentiality and security of sensitive data. The use of encryption in espionage allowed agents to relay critical information without fear of interception, safeguarding the integrity of crucial intelligence.

Cryptanalysis, the process of deciphering codes and ciphers, has become a cornerstone of intelligence operations, providing agencies with the means to decipher intercepted communications and decode encrypted messages sent by adversaries. This practice of cryptanalysis has significantly influenced the outcome of various covert operations, enhancing the ability of intelligence services to gather and interpret valuable information. The role of cryptography in espionage was not limited to securing communications; it also encompassed the development of covert communication methods, such as the use of coded signals and disguised messages to convey secret instructions and directives. These clandestine communication tactics proved essential for orchestrating covert operations and conducting infiltration missions, allowing intelligence agents to disseminate vital information while minimising the risk of exposure.

During the Cold War, cryptography continued to play a prominent role in espionage, with the two rival superpowers intensifying their efforts to develop sophisticated encryption technologies to protect their classified communications. This era was marked by an unprecedented boom in espionage activities, with cryptography serving as the linchpin for the collection and dissemination of secret intelligence.

The interaction between cryptography and intelligence highlighted the ethical and moral dilemmas associated with covert operations, raising questions about the limits of surveillance and the implications of using encryption for secrecy and manipulation. As a result, the role of cryptography in intelligence and espionage remains a fascinating topic of historical importance, highlighting the complex interconnections between technology, secrecy, and power in the realm of global espionage.

Technology Transfer: From War to Cold War

As World War II drew to a close, a different kind of war emerged: the Cold War. The technological innovations and advances in cryptography made during the war did not fall into disuse, but became essential instruments in the escalating conflict between the superpowers. The transfer of technology from the battlefield to the realm of espionage marked a profound shift in the landscape of global power dynamics. The clandestine deployment of cryptographic techniques exploited during the war gave rise to a secret arms race, where information, rather than traditional weaponry, became the primary battlefield. The United States and the Soviet Union sought to exploit the secrets of wartime cryptology to gain an advantage over each other in the complex game of intelligence gathering and surveillance.

The alliance between the Allies during the war gave way to suspicion and animosity in the post-war period, leading to the appropriation of technologies that had been shared in a spirit of common interest. This transfer of cryptographic expertise and equipment changed the nature of international relations, leading to the development of sophisticated systems dedicated to intercepting, decrypting and encrypting sensitive communications. The expertise gained during the war laid the foundation for the creation of vast intelligence agencies and surveillance networks, underpinning the strategies of the CIA and KGB, which played a decisive role in shaping history during the Cold War period.

The proliferation of cryptological knowledge and practices was not limited to government agencies. As the private sector capitalised on opportunities in the burgeoning field of cryptography, a new dimension of competition and innovation emerged. The involvement of defence contractors, research institutes, and universities in the evolution of cryptographic technologies blurred the lines between military and civilian interests, giving rise to a complex web of collaboration and rivalry. The repercussions of this diffusion extended beyond the confines of national security and permeated various facets of society, from finance and commerce to the fundamental issue of privacy. The interaction between technology transfer and its broader societal impacts led to a paradigm shift, with tools of warfare finding new applications in the realm of information management and control.

Essentially, the transition of cryptographic technology from theatres of war to the corridors of political intrigue sowed the seeds of a new era full of ambiguity and tension. The legacy of this transfer reverberates through the annals of history, echoing with the lingering echoes of secrecy, surveillance and the precarious balance of power.

Ethical and moral implications of cryptography

The evolution of cryptography in wartime has undeniably redefined the ethical and moral landscape of military intelligence and national security. As cryptanalysis techniques advanced and clandestine operations became more sophisticated, the ethical implications of using encrypted communications and decryption efforts came to the forefront of strategic decision-making. The ethical discourse encompasses a myriad of considerations, including issues of privacy, informed consent, data protection, and the implications of exploiting intercepted enemy communications. Throughout history, cryptographic activities have posed a significant challenge: balancing the need to secure classified information with the broader ethical responsibility for transparency and accountability. The main dilemma was determining the extent to which encrypted communications could be intercepted, decrypted, and exploited without compromising fundamental ethical standards and human rights.

The orchestration of covert intelligence operations and the dissemination of sensitive cryptographic intelligence raised critical questions about the ethical limits of government surveillance, individual freedoms, and the impact on civilian populations. Another crucial dimension of the ethical discourse revolved around the dual nature of advances in cryptography. While cryptographic breakthroughs have undoubtedly strengthened national security and intelligence operations, they have also fuelled debates about their potential misuse and implications for non-combatant populations. Deliberations on the ethical use of cryptology have therefore focused on the complex interaction between technological innovation, military imperatives, and the paramount pursuit of moral responsibility in the conduct of war.

The moral dimensions of cryptography extended beyond immediate conflict scenarios, permeating post-war repercussions and geopolitical dynamics. The fallout from the development of cryptographic methodologies during wartime spawned enduring ethical dilemmas, particularly during the ensuing Cold War period. This period was marked by an intensification of espionage activities worldwide, with cryptographic practices playing a critical role in the delicate balance of power, prompting profound ethical reflection on the ramifications of prolonged secrecy and subterfuge. Consequently, the ethical and moral implications of cryptological activities not only transcended the demands of wartime, but also had profound resonance across international relations, governance, and individual rights. The analysis of the ethical conundrums of cryptography is a poignant testament to the perpetual dialectic between technological progress, ethical stewardship, and the imperatives of safeguarding societies amid the vicissitudes of geopolitical conflict.

Post-war advances: a prelude to modern cryptology

After the end of the Second World War, the field of cryptology underwent a significant evolution towards modernisation and innovation. The post-war period was marked by remarkable advances in cryptographic techniques and technologies, paving the way for the evolution of modern cryptology. This period marked a fundamental transition from traditional encryption methods to the development of sophisticated algorithms and computational approaches. One of the key developments during this period was the emergence of electronic computing systems, which revolutionised the way cryptographic algorithms were designed and implemented. The use of these advanced technologies enabled cryptanalysts and decryptors to explore complex codes and algorithms with unprecedented accuracy and efficiency.

The adoption of mathematical concepts and statistical analysis in the field of cryptanalysis paved the way for new breakthroughs in the deciphering of encrypted communications. The creation of research institutions and academic programmes specialising in cryptology has played a crucial role in promoting intellectual exchange and interdisciplinary collaboration. This interdisciplinary approach has led to the integration of various fields such as mathematics, computer science and information theory, greatly enhancing the analytical capabilities and theoretical foundations of cryptological studies. With the escalation of the Cold War, the strategic importance of cryptography grew exponentially, prompting world powers to invest more in cryptology research and development. This era saw the continuous refinement of cryptographic protocols, key management systems, and authentication mechanisms, laying the foundation for contemporary cybersecurity practices.

The proliferation of digital communication networks highlighted the critical importance of secure transmission and data protection, catalysing the formulation of robust encryption standards and methodologies. Furthermore, the evolving threat landscape fostered the continuous evolution of cryptological strategies and defensive measures, prompting the conceptualisation of cryptographic frameworks resistant to emerging adversarial tactics and threats. The post-war period thus represents a transformative era in the history of cryptology, heralding the dawn of modern cryptographic paradigms and serving as a catalyst for the proliferation of cryptographic applications in various fields, including the military, intelligence, finance, and cybersecurity.

Conclusion: Secrecy and the balance of power

The relationship between secrecy and the balance of power in wartime and beyond is a complex and multifaceted topic that continues to shape global dynamics. The impact of cryptology on the outcome of World War II and subsequent geopolitical events cannot be overstated. The ability to protect sensitive information on site and intercept enemy communications played a decisive role in the course of history, underscoring the importance of maintaining superior capabilities in the field of cryptography. The post-war period was marked by a significant transition, with technological advances and the changing nature of conflict stimulating the development of modern cryptology. It became clear that nations and intelligence agencies around the world recognised the need to secure their communications and data from external threats, leading to an era of secret competition and surveillance.

As cryptographic methods evolved, the ethical and moral implications related to privacy, surveillance, and the risk of abuse of power came to the fore. The delicate balance between national security concerns and individual rights to privacy became a central point of contemporary discourse, with ongoing debates about the trade-offs between security measures and civil liberties. The proliferation of digital communications and the interconnectedness of modern societies have further reinforced the importance of cryptography in protecting critical information from malicious actors, underscoring its enduring relevance in the digital age.

Furthermore, the global landscape continues to witness the deployment of advanced encryption algorithms and the emergence of quantum cryptography, posing new challenges and opening new possibilities in the field of information security and intelligence gathering. Ultimately, the legacy of wartime cryptology remains a testament to the enduring influence of secrecy on the balance of power, offering valuable insight into the complex interplay between technological innovation, ethical considerations, and the preservation of national interests.

11
The Invisible Hand
Nazi technology and the Cold War

The continuation of war by other means

The end of the Second World War marked a turning point in global dynamics, with the victors and vanquished rapidly shifting from open conflict to the covert struggle of the Cold War. While the cessation of hostilities marked the beginning of an era of tense diplomacy, ideological antagonism and rivalry between the superpowers, it also heralded a new phase of competition: that of scientific innovation, technological prowess and strategic espionage. The post-war landscape saw the formidable array of wartime inventions and expertise reassigned to clandestine operations, defence reinforcement and ideological research. This period was marked by a profound reshaping of technology industries, military establishments and intelligence agencies, which pivoted to confront an elusive and indistinct adversary. The echoes of war resonated in the form of industrial espionage, the transfer of expertise, and the use of captured enemy assets to gain a competitive advantage in the Cold War arena. The emergence of a parallel front—a silent and invisible theatre of subterfuge and ingenuity—was fuelled by the legacy of scientific breakthroughs from the Second World War.

The intertwining of technological advances and ideological battlefields resulted in a global race to exploit the most powerful innovations for strategic advantage, transforming warfare into a multidimensional contest of intellect, creativity, and mobilisation. As such, it is imperative to explore the genesis and trajectory of this transition from World War II technological development to its applications during the

Cold War to understand the complex interaction between history, ideology, and progress. The interwoven tapestry of scientific development and political ideology during this period not only redefined the nature of warfare, but also catalysed unprecedented advances that continue to shape our world today.

Technological Espionage: East-West Technology Transfer

The period following the Second World War was marked by intensified technological espionage between East and West. In the midst of the nascent Cold War, the Soviet Union and the United States sought to gain an advantage by acquiring advanced technologies through clandestine means. This strategic quest for innovation led to the clandestine transfer of scientific knowledge, research findings, and even captured Nazi scientists from one hemisphere to the other. The complex network of espionage operations and clandestine information exchanges significantly shaped the post-war geopolitical landscape. The two superpowers engaged in unconventional wars where scientific knowledge and technological advances were coveted as essential tools for achieving strategic dominance. Fearing they would fall behind in the global arms race, both sides used vast networks of spies, informants, and intelligence agents to gather critical data on advances in aerospace engineering, nuclear technology, communications systems, and much more. The unprecedented level of secrecy surrounding these operations heightened tensions between East and West, underscoring the transformative impact of technological espionage on international relations. The ramifications of this era continue to reverberate in contemporary discussions of national security and the ethics of technological appropriation.

The role of captured Nazi scientists in Soviet and American R&D

After the end of the Second World War, the fate of captured German scientists became a crucial factor in the post-war power dynamics between the United States and the Soviet Union. The victorious Allied forces quickly launched Operation Paperclip, a secret endeavour to recruit prominent scientists, engineers, and technicians from the defeated Third Reich. These individuals possessed invaluable knowledge and expertise in fields such as rocketry, aeronautics, and weaponry, which were deemed essential to the rapidly evolving military strategies of the Cold War. The United States, under the leadership of agencies such as the Office of Strategic Services (OSS) and later the Central Intelligence Agency (CIA), launched a vast programme to identify, transport and assimilate the best German scientists into their own research and development infrastructure. One of the most notable figures in this initiative was Wernher von Braun, the famous V-2 rocket specialist, who would later play a key role in the American space programme.

Similarly, the Soviet Union seized the opportunity to acquire German scientific expertise through initiatives mirroring American efforts. Through operations such as Osoaviakhim and Alsos, the Soviet Union aggressively sought out

German scientists and resources to advance its own technological and military capabilities, laying the groundwork for the future space race and advances in nuclear engineering. The impact of these captured German scientists on the respective R&D initiatives of the United States and the Soviet Union cannot be overstated. Their contributions not only accelerated the development of advanced weapons systems and aerospace technologies, but also reshaped the geopolitical landscape by strengthening the strategic capabilities of both superpowers.

The fusion of German expertise with the resources and ambitions of the United States and the USSR paved the way for unprecedented advances in fields such as missile technology, aviation, and intelligence operations. However, the involvement of former Nazi scientists in the post-war activities of the United States and the USSR raises complex ethical and ideological questions. While their knowledge undoubtedly enabled technological advances, it also symbolised morally questionable collaboration with individuals complicit in the atrocities committed by the Nazi regime. This dilemma continues to spark debate about the ethical compromises made in the pursuit of scientific and military supremacy during the early years of the Cold War.

In retrospect, the imprint of captured Nazi scientists on the Cold War era demonstrates the complex interplay between scientific prowess, political machinations, and ethical considerations. Their influence reverberated through history, shaping the trajectory of global power dynamics and technological progress in ways that continue to resonate in contemporary discussions about the legacy of war and innovation.

The Rocket and the Intercontinental Ballistic Missile Arms Race

During the Cold War, rockets played a decisive role in shaping the geopolitical landscape. The development of long-range missile technology led to a fierce arms race between the United States and the Soviet Union, culminating in the deployment of intercontinental ballistic missiles (ICBMs). Here, we examine the central role of rockets in the escalation of tensions and strategic manoeuvring between the two superpowers.

In the aftermath of the Second World War, the United States and the Soviet Union actively sought ways to exploit the expertise of captured Nazi scientists, particularly those with knowledge of advanced rocket propulsion systems. Under the leadership of renowned figures such as Wernher von Braun and Sergei Korolev, these nations channelled their resources into developing long-range ballistic missiles capable of carrying nuclear warheads from one continent to another.

The successful launch of the Soviet R-7 Semyorka, the world's first ICBM, in 1957 sent shockwaves around the world and marked a significant shift in the dynamics of warfare and deterrence. In response, the United States accelerated its efforts to perfect missile technology, leading to the development and deployment of the Atlas and Titan series of intercontinental ballistic missiles (ICBMs), thereby establishing parity in capabilities. The race to miniaturise nuclear warheads for launch on these ballistic missiles then gave rise to an unprecedented pace of technological progress. The quest for greater range, but also greater accuracy, reliability and payload capacity stimulated research and innovation in fields ranging from materials science and aerodynamics to guidance and control systems.

The strategic importance of ICBMs extended beyond their destructive potential, as they profoundly influenced global politics and military doctrines. The concept of mutually assured destruction (MAD) emerged, underpinned by the credible threat of devastating retaliation in the event of a nuclear strike. These intercontinental missile arsenals thus became the cornerstone of American and Soviet deterrence strategies, influencing diplomatic negotiations and crisis management tactics.

The constant evolution of missile technology and the mutual need to counter the adversary's advantage fostered a perpetual cycle of progress and countermeasure development. This perpetuated a delicate balance of power that greatly influenced international relations and global security for decades.

In short, the arms race surrounding rockets and intercontinental ballistic missiles during the Cold War not only illustrated the pinnacle of technological prowess, it also highlighted the frightening spectre of catastrophic conflict. The legacy of this period resonates in current efforts to manage and mitigate the proliferation of advanced missile systems, underscoring the enduring impact of rocketry on international affairs and strategic stability.

Submarine Warfare Technology and Naval Strategies of the Cold War

The Cold War was characterised by a tense naval standoff between the superpowers, with submarine warfare technology and strategies playing a critical role in the dynamics of the era. Both the United States and the Soviet Union invested heavily in the development of advanced submarine fleets to maintain their strategic advantage and deterrence capabilities. Submarines became an essential tool for covert intelligence gathering and strategic deterrence during the Cold War. The development of nuclear-powered submarines, such as the USS Nautilus and the Soviet K-3 Leninskiy Komsomol, revolutionised submarine warfare by allowing submarines to operate underwater for long periods of time without having to resurface.

These advances transformed submarines from simple warships into undetectable platforms from which they could launch ballistic missiles, extending the range of nuclear arsenals beneath the waves. The deployment of submarine-launched ballistic missile (SLBM) submarines armed with intercontinental ballistic missiles (ICBMs) was a major turning point in submarine warfare. This development posed a new challenge to Cold War strategists, as SSBNs patrolled undetected beneath the ocean's surface, demonstrating unprecedented stealth and survivability. The constant cat-and-mouse game between submarine hunters and their elusive prey sparked a technological arms race in the field of anti-submarine warfare, which spawned innovative sonar technologies and sophisticated anti-submarine warfare tactics. Naval strategies evolved to integrate submarine operations into broader geopolitical considerations. Submarine-launched ballistic missiles (SLBMs) were a key element of the doctrine of mutually assured destruction (MAD), reinforcing the credibility of nuclear deterrence by providing a survivable second-strike capability.

The ability of submarines to remain hidden in distant waters provided a powerful means of projection, power projection, and crisis management, shaping maritime strategies in global hot spots. To counter the growing threat from Soviet submarines, the US Navy pursued the forward deployment of hunter-killer submarines, tasked with locating and neutralising potential enemy submarines. At the same time, the Soviet Union sought to develop its submarine fleet with an emphasis on quiet propulsion systems and improved soundproofing techniques to counter Western detection capabilities.

In short, Cold War submarine warfare technology and naval strategies not only illustrate the technological prowess of the superpowers, but also highlight the essential role that submarine warfare played in shaping international relations. The clandestine nature of submarine operations and the constant quest for technological superiority contributed to a dramatic and often overlooked aspect of the maritime dimension of the Cold War.

The influence of Nazi communication systems on Cold War tactics

The aftermath of the Second World War marked the beginning of an era defined by rivalry between the United States and the Soviet Union: the Cold War. As the two superpowers sought technological superiority to gain strategic advantage, the influence of Nazi communication systems emerged as an important factor in the evolution of Cold War tactics. The Nazis had developed advanced communication technologies, including encrypted radio transmissions and secure telecommunication systems, which had immense potential for military application. After the war, these technologies became coveted assets, fuelling a clandestine race to acquire and adapt them for modern warfare. One of the most notable influences was the Enigma machine, an encryption device used by the Nazis to encrypt their communications. Its cryptographic complexity posed a significant challenge to Allied code breakers during the Second World War and remained a subject of intrigue in the post-war period. The United States and the Soviet Union strove to exploit the principles underlying the Enigma machine for their own intelligence operations.

As a result, advances in cryptanalysis and the development of sophisticated encryption methods were spurred by the quest to decode and emulate the security mechanisms used by the Nazis. The legacy of Nazi communication systems manifested itself in the formulation of secure and resilient communication networks designed to withstand the rigours of modern warfare. The complex and robust infrastructure established by the Nazis served as a model for the establishment of secure military communications capable of meeting the challenges posed by electronic warfare and espionage.

The integration of wireless communication technology, as demonstrated by advances in portable radio devices and encrypted voice transmissions, is a testament to the lasting impact of Nazi innovation on the evolution of communication systems during the Cold War era. Furthermore, the use of signals intelligence, or SIGINT, derived from Nazi expertise in intercepting and deciphering enemy communications, became the cornerstone of Cold War espionage and reconnaissance operations. Electronic countermeasures, jamming and deception strategies, all based on an understanding of Nazi communication methods, became essential tools in the arsenal of Western and Eastern bloc intelligence agencies. The knowledge gained from studying Nazi communication practices helped refine SIGINT capabilities to monitor adversaries and protect information vital to national security.

In short, the pervasive influence of Nazi communication systems on Cold War tactics underscores the lasting impact of wartime technological developments on the subsequent geopolitical landscape. By exploiting and adapting the innovations of the past, the nations engaged in the Cold War perpetuated a legacy of ingenuity encompassing cryptography, wireless communication, and signals intelligence, thereby reshaping the contours of modern warfare and espionage.

Advances in Aviation: From the ME 262 to the SR-71 Blackbird

As the world transitioned from the devastation of World War II to the tensions of the Cold War, Nazi aviation technology continued to influence and shape the trajectory of aerial warfare. The Messerschmitt Me 262, the world's first operational jet fighter, revolutionised aerial combat with its formidable speed and agility. Its impact on post-war aviation was profound, as both the United States and the Soviet

Union sought to capitalise on its technological advances. This era saw a race to develop faster, more manoeuvrable and undetectable aircraft, reflecting the evolving nature of aerial warfare. The following decades saw the emergence of advanced aircraft that pushed the boundaries of flight and reconnaissance. In the United States, the development of the Lockheed A-12, the precursor to the iconic SR-71 Blackbird, represented a leap forward in high-altitude, high-speed reconnaissance capabilities. Its sleek design and advanced engineering allowed it to soar at unprecedented speeds, making it virtually untouchable by enemy defences. The SR-71 Blackbird became synonymous with stealth and precision, embodying the pinnacle of Cold War aeronautical technology.

Simultaneously, on the other side of the Iron Curtain, the Soviet Union pursued its own advances in aviation to counter the technological prowess of the West. The Sukhoi Su-27 Flanker, an agile and versatile fighter aircraft, exemplified the Soviet commitment to developing competitive aircraft. With air superiority becoming a key objective for both superpowers, the race for dominance of the skies intensified, leading to a series of technological breakthroughs that redefined the parameters of aerial warfare. The legacy of Nazi aviation technology extended far beyond the conflicts of the time, influencing the design principles and strategic perspectives of subsequent generations of aircraft. The knowledge and expertise gained from wartime adversaries laid the foundation for unprecedented advances in aerospace engineering, shaping the ideological and tactical landscape of the Cold War. The interaction between historical legacies and contemporary innovations underscored the central role of aviation in the struggle for global supremacy. From the pioneering spirit of the Me 262 to the stealth and precision of the SR-71 Blackbird, the journey of aviation progress marked an essential chapter in the annals of military and technological history.

Psychological Warfare: Propaganda Techniques Redeveloped

During the Cold War era, the resurgence of psychological warfare and propaganda became an essential element in the geopolitical struggle between the United States and the Soviet Union. Both superpowers quickly recognised the power to influence public opinion and shape perceptions both domestically and internationally. The use of propaganda techniques evolved to exploit modern media, culture, and communication channels to achieve strategic objectives. Learning from Nazi Germany's effective use of propaganda, both sides sought to exploit similar tactics to manipulate public opinion and advance their ideological agendas.

A striking aspect of this evolution was the integration of technology into propaganda efforts. The media played a crucial role in disseminating carefully crafted messages to target audiences worldwide. State-of-the-art printing presses, radio broadcasting, and later television and film were all enlisted to wage information warfare. The emergence of innovative communication technologies enabled the rapid transmission of tailored narratives to influence international opinion and weaken adversaries.

Furthermore, the launch of clandestine programmes designed to infiltrate and manipulate foreign media marked a new chapter in psychological warfare. These operations rely on sophisticated techniques to spread disinformation, cause unrest and discredit rival factions while supporting friendly alliances. The clandestine dissemination of false narratives and falsified documents has served as a tool to shape public discourse, erode trust in institutions and influence the course of events in target countries. This strategic game of narrative manipulation had profound repercussions on the balance of power during the Cold War. In addition to traditional propaganda, cultural initiatives and psychological operations were integrated to serve the objectives of both superpowers. Academic exchanges, artistic sponsorships and cultural events were used to project a favourable image and promote ideological superiority. Simultaneously, intelligence services sought to subvert the cultural exports and movements of the opposing side to weaken unity and provoke internal discord. This renaissance of psychological warfare reshaped the ideological battlefield, blurring the lines between conventional warfare and information warfare. The legacy of propaganda techniques redeveloped during the Cold War attests to the enduring influence of psychological operations on global politics and public perception.

Economic Warfare: Leveraging Industrial Innovations

Economic warfare during the Cold War took on a new dimension with the strategic exploitation of industrial innovations derived from Nazi technology. As nations vied for supremacy in a global power struggle, the application of cutting-edge advances in various industries became instrumental in gaining and maintaining economic dominance. The impact of industrial innovations derived from Nazi technology shaped the geopolitical landscape and revolutionised the nature of economic competition.

The exploitation of captured Nazi scientists and their innovative expertise ushered in a new era where technological superiority was directly linked to economic power. The infusion of German expertise into sectors such as aerospace, engineering, and medicine fostered the development of revolutionary technologies that would redefine entire industries. The technological spin-offs resulting from this infusion triggered a period of rapid progress and innovation, fundamentally altering the international economic hierarchy.

Furthermore, the effective use of industrial innovations proved to be a formidable tool for dismantling the economic strength of adversaries. Nations strategically deployed these advances to bolster their own industries while diluting the economic capabilities of rival powers. This manipulation of technological prowess not only disrupted traditional economic paradigms, but also exerted considerable influence on diplomatic and military strategies, reinforcing the interdependence of economic and geopolitical power in the post-war world.

The use of industrial innovations derived from Nazi technology led to the establishment of symbiotic relationships between industry and government, blurring the lines between private enterprise and national interest. This fusion resulted in the formation of powerful economic-military complexes that exerted unprecedented influence over domestic and international affairs. The indelible mark of Nazi technological advances on this symbiotic relationship prompted a reassessment of the role of industrial institutions in the fabric of nation-state dynamics, galvanising a shift in the global economic order.

In short, the strategic deployment of industrial innovations derived from Nazi technology constitutes an essential chapter in the history of economic warfare during the Cold War. The use of these advances altered the trajectory of global economic competition, redefined the parameters of national power, and catalysed a metamorphosis in the relationship between technology, industry, and the art of governance. Understanding the complex interaction between technological prowess and economic supremacy provides insight into the multifaceted geopolitics of the Cold War and highlights the enduring legacy of Nazi technology on the global economic stage.

Conclusion: The Dual Legacy of Nazi Technology in the Context of the Cold War

The intertwined legacy of Nazi technology in the context of the Cold War is complex and multifaceted. As the world transitioned from the devastation of World War II to the political and military stalemate of the Cold War, the remnants of Nazi technological prowess continued to shape the trajectory of global power dynamics. This chapter has peeled back the layers of history to reveal how industrial innovations born from the dark crucible of Nazi Germany left an indelible mark on the landscape of conflict, espionage, and competition between the United States and the Soviet Union. With a dual legacy that profoundly influenced the technological, military, and political developments of the period, the importance of Nazi technology cannot be overstated.

On the one hand, the Cold War era saw the continuation and evolution of the economic warfare strategies initiated by Nazi Germany. Technologies such as advanced weaponry, aerospace engineering and communications systems were repurposed and exploited to gain strategic advantages and assert dominance in the new bipolar world order. The fruits of Nazi industrial innovations were reaped by both the Western and Eastern blocs, significantly influencing the arms race, naval strategies, aviation advances, and intelligence operations.

On the other hand, the intertwined legacy also propagated an ethical and moral dilemma. Indeed, the use of technology developed under the Nazi regime raised difficult questions about complicity and responsibility. The contribution of former Nazi scientists and engineers to the space race, missile programmes, and intelligence agencies highlighted the controversial nature of exploiting knowledge derived from an oppressive and tyrannical regime. Furthermore, the psychological impact of reusing Nazi propaganda techniques and exploiting industrial innovations added a new layer of complexity to this dual legacy. As the Cold War progressed, the ethical implications of adopting these technologies became increasingly pronounced, fuelling debates and controversies that reverberated through the corridors of power.

Ultimately, the dual legacy of Nazi technology during the Cold War highlights the complex interaction between innovation, ethics and geopolitics. It serves as a stark reminder of the enduring influence of history on the present and the future, compelling us to confront the complexities and consequences of technological legacies shaped by tumultuous historical periods.

12
The Mysteries Of Area 51
Fact or Fiction?

The History of Area 51

Located deep in the Nevada desert, Area 51 has been shrouded in mystery and controversy since its creation. The history of this enigmatic place has captured the public's imagination, giving rise to a multitude of myths and legends that have spanned decades. To truly understand the intrigue surrounding Area 51, it is imperative to delve into the historical origins and evolutionary trajectory of the myths and conspiracies that have become synonymous with its name. The idea of a secret government facility engaged in clandestine activities has fuelled speculation and fascination, perpetuating the enigma surrounding this highly confidential establishment.

Let us not overlook the role of popular culture, which has greatly contributed to the mystique of Area 51 by portraying it as a centre for extraterrestrial encounters, top-secret military experiments, and advanced technological innovations beyond human comprehension. These narratives, whether fuelled by sincere curiosity or fervent scepticism, have shaped the cultural landscape and collective consciousness, manifesting themselves in a plethora of artistic expressions, literature and entertainment media. As we embark on an exploration of the origins and evolution of these persistent myths, it becomes clear that Area 51 occupies a unique position at the intersection of speculation, intrigue, and government secrecy, leaving an indelible mark on the tapestry of modern folklore and conspiracy theories.

Historical Context and Creation

The enigmatic aura surrounding Area 51 invariably leads our exploration into the corridors of history, searching for the origins and purpose of this clandestine realm. Delving into the historical context, it becomes evident that the creation of Area 51 was an essential response to the global atmosphere that prevailed at the time of its conception. A product of the Cold War era, Area 51 came into being as part of the US government's quest for supremacy in weapons technology and strategic defence. Its creation was driven by the urgent need to conduct top-secret research and testing on advanced aircraft and weapons.

With the world on the brink of nuclear conflict and espionage permeating international relations, Area 51 emerged as a secret crucible for technological innovation and tactical prowess. The enigmatic nature of its creation and expansion illustrates the confluence of political tensions, military ambition, and scientific ingenuity that defined the Cold War era. The mysteries surrounding its formation invite speculation and intrigue, adding to the allure of this mysterious domain.

The creation of Area 51 has significance that extends beyond the realm of military activities. It symbolises the pinnacle of classified research and development, embodying the clandestine activities of multiple government agencies converging within its borders. The complex web of connections linking Area 51 to broader geopolitical strategies contributes to its enigmatic status and fosters an evocative narrative that resonates with historical significance. Unravelling the historical tapestry of Area 51 provides profound insights into the complex interplay between national security imperatives and scientific progress, highlighting the complex role it has played in shaping the world order over the past century.

Alleged extraterrestrial encounters

The notion of extraterrestrial encounters in the vicinity of Area 51 has become deeply intertwined with its mystical and enigmatic character. Reports of unidentified flying objects (UFOs) and alleged encounters with beings from other worlds have perpetuated the legend surrounding this clandestine facility. Although scepticism may cast doubt on the veracity of these reports, the appeal and fascination with the idea of extraterrestrial visits persists.

Conspiracy theories and testimonies from alleged witnesses continue to fuel speculation about possible government cover-ups and secret interactions with advanced beings. The alleged Roswell incident in 1947, in which a UFO reportedly crashed near Roswell, New Mexico, has been closely linked to Area 51, further amplifying its association with extraterrestrial phenomena. Despite official explanations attributing these accounts to experimental aircraft or natural phenomena, the allure of extraterrestrial encounters remains a central element of Area 51 folklore.

Unidentified aerial phenomena (UAP) documented by military and civilian personnel have helped fuel discourse about alleged extraterrestrial activities. Detailed eyewitness accounts and video evidence have captured public interest, leading to calls for transparency and disclosure from government authorities. However, the complex interplay of secrecy, national security, and public curiosity continues to shroud alleged extraterrestrial encounters in mystery and controversy.

Exploring the psychological and sociocultural dimensions of alleged extraterrestrial encounters reveals their profound impact on the popular imagination and collective consciousness. These narratives often transcend the boundaries of conventional reasoning, inviting individuals to contemplate the vast unknown and question humanity's place in the cosmos. As we delve into the complexities of extraterrestrial claims associated with Area 51, it is imperative to navigate the intersection of scientific inquiry, speculative conjecture, and the eternal aspiration for answers that lie beyond our earthly realm.

Technology Beyond the Horizon

Emerging from the veil of secrecy that shrouds Area 51, discussions surrounding the alleged technological advances and experiments conducted within its enigmatic confines have perpetuated speculation and intrigue.

The notion of 'technology beyond the horizon' encapsulates the prevailing belief that the innovations developed at this clandestine facility transcend the limits of conventional understanding and herald a new era of scientific achievement. Rumours and accounts from alleged insiders have perpetuated claims that radical propulsion systems, energy sources, and material compositions far exceed current human knowledge. These alleged breakthroughs have also been linked to sightings of unidentified aerial phenomena (UAP) in the vicinity of Area 51, contributing to the atmosphere of enigma and mysticism surrounding this secret location. The speculative nature of these claims has sparked heated debate among conspiracy theorists, sceptics, and researchers, each striving to discern the veracity of these extraordinary claims.

The dichotomy between the US government's presentation as a bastion of transparency and the impenetrable veil of secrecy surrounding the activities of Area 51 only intensifies public curiosity about what lies beyond the established limits of technological progress. The implications of these advances, if proven true, are profound and could revolutionise fields ranging from aerospace engineering to materials science and beyond. Here, we seek to delve deeper into the fervent discourse and provide a comprehensive examination of the technological phenomena that ostensibly exist beyond the perceived limits of human innovation.

Nazi scientists and classified experiments

The post-war period saw a significant exodus of German scientists, engineers, and researchers to various parts of the world, including the United States, as part of Operation Paperclip. Among this influx of scientific minds were individuals who had been associated with Nazi Germany's technological and scientific efforts. Their expertise in fields such as rocketry, aerospace engineering, and physics made them attractive assets for the burgeoning intelligence and defence programmes of the Cold War era. However, the recruitment of these individuals raised ethical and moral questions, given their association with a regime responsible for heinous atrocities. We therefore note the complex confluence of historical circumstances, scientific acumen, and geopolitical imperatives that led to the participation of former Nazi scientists in classified experiments.

The assimilation of these experts into American scientific institutions and military research laboratories marked a controversial but influential chapter in the annals of technological development. Their contributions, while significant, also prompt reflection on the ethical considerations surrounding scientific collaboration and the consequences of aligning with individuals linked to reprehensible actions. By closely examining this complex interplay between knowledge acquisition, national security imperatives, and moral deliberation, we arrive at a nuanced understanding of the profound impact these scientists had on classified experiments and the scientific landscape as a whole.

Government secrecy and public speculation

Government secrecy surrounding Area 51 has been the subject of fervent debate and public speculation for decades. The veil of secrecy surrounding this enigmatic facility has fuelled a plethora of theories, ranging from mundane military operations to extravagant claims of extraterrestrial involvement. This veil of secrecy has created a vacuum in which conspiracy theories have sprouted and flourished, often born from the fertile imaginations of individuals seeking to unravel the mysteries surrounding Area 51. The lack of transparent information and official recognition from government authorities has contributed to the proliferation of wild speculation about the activities taking place within the confines of Area 51. The lack of concrete data has led to much conjecture, with some claiming that the government is concealing evidence of extraterrestrial contact or technological advances far beyond the realm of conventional understanding.

The clandestine nature of Area 51 has fuelled suspicion and speculation about the true nature of the experiments being conducted there. In the absence of clear and open communication from government agencies, the public has been left to construct its own narratives to fill the information void. The result is an atmosphere of scepticism and mistrust, which casts doubt on the veracity of official statements and fuels conjecture. As a result, the perceived intent of government secrecy has become a subject of intense speculation in popular discourse, giving rise to questions about the ethical implications of withholding information from the public.

Many advocates of transparency argue that the concealment of Area 51's activities constitutes a violation of democratic principles, citing the need for accountability and openness in matters of national importance. Conversely, proponents of government discretion argue for the need to maintain confidential operations for reasons of national security, asserting that certain knowledge must remain shielded from public scrutiny to safeguard strategic interests. The conflict between public speculation and government discretion has perpetuated a lasting cycle of intrigue and mysticism surrounding Area 51. The persistent lack of conclusive evidence has only perpetuated the enigma, ensuring that Area 51 remains a focal point of tantalising speculation and ongoing controversy.

Testimonials and whistleblowers

The enigmatic aura surrounding Area 51 has spawned a plethora of testimonials and whistleblower accounts, adding new layers of intrigue to an already puzzling narrative. Numerous individuals have come forward over the years, claiming to have participated in or witnessed first-hand clandestine activities within the confines of this secret military facility. These testimonies often involve sightings of unconventional aircraft, anomalous phenomena, and alleged encounters with otherworldly entities. Several whistleblowers, believed to be former employees or insiders at the enigmatic facility, have sought to expose alleged government cover-ups and the truth about what they believe to be a series of unconventional experiments and advanced technological developments. While some of these claims have been met with scepticism and scrutiny, they nonetheless contribute to the complex tapestry of beliefs and suspicions surrounding Area 51.

One particularly notable account is that of Bob Lazar, who drew attention to his claims about his alleged work at a clandestine base near Area 51, where he claims to have encountered extraterrestrials and researched their technology. Despite widespread scepticism, Lazar's allegations have remained central to the debate surrounding Area 51. Other testimonies offer a wide range of accounts, from alleged sightings of extraterrestrials and UFO phenomena to descriptions of highly advanced aerial craft that defy conventional understanding.

These accounts have not only fuelled conspiracy theories,

but have also given rise to numerous sensationalist speculations and interpretations in popular culture, leading to a proliferation of diverse narratives and interpretations regarding the true nature of Area 51 and the activities associated with it. It is essential to approach these testimonies and allegations with discernment, recognising the complex interplay of factors such as misinformation, misinterpretation, psychological influences, and potential motives for disseminating these narratives. By critically evaluating these narratives within the broader context of historical and socio-cultural factors, we can strive to gain a more complete understanding of the enduring fascination and mystery surrounding Area 51.

Analysis of declassified documents

Upon entering the enigmatic realm of Area 51, one inevitably encounters a labyrinth of clandestine activities, shrouded in secrecy and speculation. While eyewitness accounts and whistleblowers have shed light on certain aspects of this clandestine location, the analysis of declassified documents is an essential key to unravelling its mysteries. These invaluable documents, emerging from a veil of confidentiality, offer a rare glimpse into the historical tapestry of Area 51 operations. From the Cold War era to the present day, these documents are artefacts of innovation, experimentation and government programmes. A meticulous examination of these declassified documents reveals a multidimensional narrative that invites readers to question the veracity of prevailing theories and discover the truth behind the mystique. The unveiling of the documentation surrounding Area 51 leads to an intellectual expedition, punctuated by reports on advanced aircraft testing, the development of experimental weapons and secret reconnaissance missions. Furthermore, it highlights the interaction between geopolitical tensions, defence strategies and technological advances at pivotal moments in history. The intrigue intensifies when confronted with redacted segments, arousing curiosity about the sensitive information hidden in the annals of classified history.

Close examination of these documents forces us to explore the psychological and sociological implications inherent in maintaining such a veil of secrecy. It invites us to contemplate the ethical dimensions of government opacity and the ramifications of selectively concealing knowledge in the public domain. While casting doubt on the alleged extraterrestrial connection to Area 51, analysis of the declassified documents fuels evidence-based discourse, prompting readers to critically deconstruct prevailing myths and conjecture. It encourages a methodical dissection of evidence, fostering scepticism while encouraging an empirical investigative mindset. Through this process, the empyreumatic allure of Area 51 is dissected, allowing readers to discern between factual accounts and sensationalist folklore. Ultimately, analysis of the declassified documents serves as a channel of enlightenment, allowing for a recalibration of perspectives and an incisive reassessment of the captivating enigma that is Area 51.

Area 51 in popular culture

The enigmatic and secretive nature of Area 51 has captured the imagination of popular culture around the world. It has been depicted and referenced in numerous films, television programmes, books, video games, and conspiracy theories. The mystique surrounding Area 51 has fuelled a variety of narratives, ranging from speculative fiction to the wildest conspiracy theories. In cinema, Area 51 has served as the setting for countless science fiction and alien films. Hollywood has perpetuated the idea that Area 51 was a centre for alien secrets, advanced technologies, and clandestine government operations. Whether in blockbusters or independent films, the allure of Area 51 continues to captivate audiences around the world.

Television shows and documentaries have helped to cement the myth of Area 51 in popular culture. Numerous television series have explored the myths and rumours surrounding the facility, often blurring the lines between reality and fiction. These portrayals contribute to the mystique and fascination that the general public has with this elusive place. Beyond traditional media, literature and online platforms are rife with stories and discussions about Area 51. Countless novels, short stories, and graphic novels have used the secrecy of Area 51 as a backdrop for thrilling narratives.

Internet forums, social media, and online communities serve as hubs for enthusiasts, sceptics, and conspiracy theorists, fuelling the conversation and contributing to the ongoing intrigue surrounding the facility. Due to its pervasive presence in popular culture, Area 51 has become an enduring symbol of mystery and speculation. Its influence extends far beyond mere entertainment, seeping into broader conversations about government transparency, technological advances, and the possibility of extraterrestrial life. The cultural impact of Area 51 is a testament to the enduring appeal of enigma and the power of human curiosity.

Conclusion: Separating Myth from Reality

As the dust settles on the enigmatic allure of Area 51, it becomes imperative to embark on a journey of discernment, separating the layers of myth from the foundation of reality. While the narratives surrounding Area 51 have woven a captivating web of speculation and intrigue, the search for truth requires careful scrutiny and critical analysis. The elusive nature of classified military installations has created an atmosphere conducive to conspiracy theories and otherworldly speculation, resulting in a pervasive cultural fascination with Area 51. However, through thorough investigation and meticulous examination of historical records, it is essential to approach the enigma of Area 51 with tempered rationality and empirical evidence. One of the key challenges in unravelling the mysteries of Area 51 is distinguishing authentic accounts from embellished folklore. Sensational descriptions of extraterrestrial encounters and advanced technological marvels must be juxtaposed with verified sources and substantiated testimonies. Delving into the annals of declassified documents and first-hand reports offers a nuanced perspective that allows one to separate fact from fiction.

Furthermore, the history of Area 51, linked to secret government activities and classified experiments, requires careful dissection of rumour and reality. The imprint of clandestine Cold War operations and the presence of pioneering scientists from diverse backgrounds cast a complex shadow over the mystique of Area 51. It is within this labyrinth that the contours of truth emerge, challenging popular narratives and inviting a sober assessment of what lies beneath the veil of secrecy.

At the same time, Area 51's enduring impact on popular culture and the media has perpetuated a powerful mix of awe and scepticism. Through film, literature and speculative discourse, Area 51 has transcended its physical boundaries and woven itself into the fabric of contemporary mythology. Recognising this pervasive influence, the process of delineating reality involves dismantling the layers of romanticism and hyperbole that have shrouded Area 51 in a cloak of enigma. Ultimately, the quest for truth regarding Area 51 requires intellectual rigour, careful scrutiny, and judicious evaluation of all available evidence. It is incumbent upon researchers and enthusiasts to chart a course through the nebulous terrain of Area 51, guided by the compass of reason and empirical inquiry. By navigating the confluence of fact and conjecture, the enigmatic legacy of Area 51 can be distilled into a narrative based on verifiable truths, providing insight into a domain that has captivated the human imagination for decades.

13
Veiled Innovations
Secret Projects

Decoding the veil

The journey into the realm of clandestine technological developments begins with an exploration of the complex web of secrecy surrounding these enigmatic projects. Throughout history, these secret innovations have played a vital role in the evolution of nations and wars. Curious minds are invited to delve into the depths of this veil, where cutting-edge engineering, strategic foresight and unwavering discretion intertwine. Decoding the veil that covers these fundamental projects is like unravelling a tapestry of intrigue, woven from the finest threads of concealment. This examination is not a mere foray into historical curiosities, but a deep dive into humanity's ability to wield knowledge as a weapon. It reveals the cunning with which powers have sought to gain the upper hand by creating and protecting cutting-edge technologies, propelling the world into uncharted territories of progress and peril.

To untangle the complexities underlying these hidden initiatives requires a keen appreciation of the underlying motives that drove their creation and the ramifications they continue to have on the fabric of our society. With each revelation, a curtain is lifted, unveiling a tapestry of ambitions, risks, and consequences that transcend simple patterns of innovation. The mystery lies not only in the discreet designs and applications, but also in the ethical shadows cast by the surreptitious efforts of governments and organisations. Throughout this intellectual odyssey, it becomes clear that decoding the veil transcends the mere unveiling of technology; it invites contemplation of humanity's penchant for concealment and the concomitant revelations that echo through the corridors of time.

The Blueprints of Secrecy: Founding Projects

The clandestine world of secret projects during World War II and the Cold War that followed represented a paradigm shift in technological innovation. Within top-secret facilities and under the cover of utmost discretion, fundamental projects were launched to push the boundaries of science and engineering. These secret plans laid the groundwork for revolutionary developments that would shape the course of history. The unveiling of the clandestine efforts of this era sheds light on the enigmatic link between technological progress and national security. The foundational projects encompassed a diverse range of activities, spanning a broad spectrum of disciplines. From advanced weaponry and aerospace systems to clandestine energy projects and secret communications technologies, the scope of these projects was as vast as it was ambitious. The genesis of these projects often stemmed from the urgent needs of war, leading to the convergence of brilliant minds in an atmosphere of absolute secrecy.

In various clandestine research centres, the brightest scientists, engineers and innovators worked in obscurity, driven by a shared mission of the utmost importance. The concealment of these fundamental projects was not simply a matter of bureaucratic protocol; it was a fundamental principle aimed at preventing essential technological advances from falling into the hands of adversaries. The strategic implications of these initiatives underscored the need for an unprecedented level of secrecy.

In this realm of clandestine creativity, plans for secrecy materialised into tangible manifestations of innovation and ingenuity. The confidential nature of these projects fostered an environment where risk-taking and bold experimentation thrived, shielded from the prying eyes of the public. The inherent challenge of balancing the imperative of secrecy with the relentless pursuit of progress defined the ethos of these fundamental projects. As such, the narrative of these secret endeavours illustrates the interface between the imperative of national defence and the relentless pursuit of technological breakthroughs. To understand the legacy of these fundamental projects is to venture into the enigmatic corridors of technological evolution and the complex tapestry of historical intrigue.

Advanced Weapons Systems: Creation and Concealment

During the tumultuous era of secret technological advances, advanced weapons systems stood as imposing monuments to human innovation and secrecy. From the inception of clandestine projects, a veil of secrecy shrouded revolutionary developments in weaponry that reshaped the landscape of warfare. The creation and concealment of these advanced weapons systems became a strategic imperative for nations engaged in global conflict. Ingenious minds leveraged their expertise to design cutting-edge weaponry, ranging from stealth aircraft and hypersonic missiles to directed-energy weapons and unmanned combat vehicles. These game-changing innovations represent the pinnacle of military technology, fuelled by clandestine research and development programmes conducted in the utmost secrecy. Orchestrating these projects required an unparalleled level of discretion and compartmentalisation, as governments and defence agencies strove to protect their revolutionary advances from prying eyes and hostile adversaries.

Within the secret enclaves of remote research facilities and testing grounds, science fiction became reality as pioneering engineers and scientists pushed the boundaries of what was thought possible, harnessing the power of classified knowledge to create unrivalled instruments of war. The complex dance between creation and concealment unfolded within the hallowed halls of cutting-edge laboratories and underground facilities, where the alchemy of ingenuity and discretion gave birth to a new breed of weapons systems destined to redefine the art of warfare.

As secrecy shrouded every milestone in the weapons' design and functionality, the world remained unaware of the magnitude of the advances that were about to tip the balance of power on the global stage. However, behind the iron curtain of classification, a symphony of progress orchestrated by exceptional minds sculpted the future battlefield with innovations designed to revolutionise tactical capabilities and strategic dominance. The legacy of these clandestine projects stands as a testament to humanity's ability to harness technological prowess in pursuit of military superiority, forever etching a chapter of mystery and wonder into the annals of history.

Biochemical activities: hidden objectives

Exploring biochemical activities during the era of secret projects reveals a labyrinth of clandestine pursuits and covert research. At the intersection of science and secrecy, clandestine facilities explored the enigmatic realms of biological warfare and medical experimentation. Hidden in the shadows, these efforts sought to harness the power of biology for offensive and defensive purposes, shaping the landscape of wartime strategies and post-war implications. The pursuit of biological weapons, characterised by their potential for large-scale devastation and undetectable deployment, drove various clandestine agencies and organisations to push the boundaries of ethics and the frontiers of science.

We examine the complex tapestry of biological research, highlighting the secret agendas that advanced biochemistry and medical science in an era of secrecy and suspicion. From the development of deadly pathogens designed to cause mass casualties to the exploitation of human subjects in unethical experiments, the tangled web of biochemical activities reveals dark chapters in history. The dichotomy between medical advances for the betterment of humanity and the weaponisation of disease for destructive purposes highlights the moral complexities of these hidden agendas. To unravel the enigma of biochemical pursuits, one must carefully examine the ethical lapses, scientific achievements, and lasting legacy of these veiled initiatives. Venturing deeper into this realm, the book explores the secret laboratories, elusive figures, and obscure narratives that illustrate the astonishing pursuits and ethical dilemmas in the clandestine world of biochemical activities.

Communications Technology: Secret Developments

In the aftermath of the Second World War, advances in communications technology took a secretive turn as different nations sought to gain a strategic advantage. Secret projects were launched to develop sophisticated communications systems with enhanced encryption capabilities and long-distance transmissions. The clandestine nature of these developments meant that they escaped public knowledge and traditional regulatory frameworks. High-level research and development teams were formed in the utmost secrecy to develop cutting-edge communications technologies. These initiatives included the creation of secure and undetectable communication channels, which played a vital role in clandestine operations and intelligence gathering.

The integration of emerging disciplines such as cryptography, signal processing, and information theory laid the foundation for unprecedented advances in covert communications. A key objective of these clandestine developments was to ensure immunity from enemy interception and decryption.

Signal obfuscation techniques, frequency hopping, and spread spectrum modulation formed the basis of secure transmission protocols, allowing encrypted messages to travel vast distances without compromise. The proliferation of electronic surveillance necessitated the development of countermeasures, leading to the creation of highly secretive anti-surveillance communication systems.

The evolution of satellite communication marked a turning point in covert technology, offering global reach for encrypted transmissions. Hidden in satellite constellations and disguised as harmless payloads, clandestine communications satellites facilitated secure, untraceable exchanges between remote operatives and headquarters. These satellites were an invaluable asset to intelligence agencies and military commands operating in regions lacking traditional communications infrastructure.

As the Cold War intensified, covert communications technologies gained prominence, with the advent of encrypted telex, secure voice transmission, and encrypted data networks. The advanced encryption standards implemented in these systems were classified at the highest level and remained hidden from public view for decades. These considerable efforts to secure means of communication underscore the critical role of information superiority in clandestine operations of the era. The enduring legacy of these secret developments in communications technology continues to resonate in contemporary intelligence and defence operations. The principles and methodologies born of these clandestine efforts have continued to shape the modern landscape of secure and resilient communications systems, ensuring that sensitive and critical information remains out of reach of adversaries.

In the shadow of clandestine operations, energy projects emerged as silent giants in the fabric of secret technological activities. These initiatives explored the realms of sustainable and advanced energy sources, shielded from public attention. The search for clandestine energy solutions was driven by the need for independence, reliability, and superiority in the global power game. Cut off from conventional oversight, these projects operated in the dark, allowing revolutionary innovations to develop beyond traditional limits and regulations.

Aerospace Innovations: Beyond the Public Eye

Exploring aerospace innovations beyond the public eye reveals a clandestine world where advances in aeronautics and space exploration were made in the utmost secrecy. Behind closed doors, aerospace engineers and scientists worked tirelessly to push the boundaries of flight and propulsion systems. Hidden from public view, experimental aircraft and spacecraft were developed using cutting-edge technologies and revolutionary concepts. These aerospace innovations, hidden from public view, were born out of a marriage of necessity and competition, as nations strove to gain the upper hand in the post-war era. Under the cloak of secrecy, technological marvels emerged and transformed the aerospace landscape. Unmanned aerial vehicles (UAVs) or drones, often associated with modern warfare, have their conceptual roots in secret projects of the past.

The development of hypersonic aircraft, capable of travelling at speeds exceeding Mach 5, remained deeply entrenched in classified programmes for decades, until recent revelations brought their existence to light. In addition to advances in aviation, secret space projects propelled humanity into uncharted territory. Lunar missions, space planes, and advanced rocket programmes have been shrouded in secrecy, each with the potential to reshape our understanding of space exploration. The veil of secrecy surrounding these aerospace innovations has not only obscured monumental achievements, but also the stories of intrepid pioneers who dared to tread the uncharted frontiers of the cosmos.

Stealth technology, essential for evading detection by radar and other forms of surveillance, has been intensively developed as part of clandestine aerospace projects. Radical airframe designs, new composite materials, and sophisticated radar-absorbing coatings are among the revolutionary innovations hidden from public view. These technologies have resulted in the creation of iconic stealth aircraft, ushering in a new era in aerial warfare while revolutionising the principles of aircraft design and manufacture.

As revelations surface, shedding light on former secret aerospace projects, a world of ingenuity and dedication hidden from public view becomes apparent. The legacy of these secret endeavours remains woven into the fabric of contemporary aerospace achievements, serving as a testament to human perseverance, innovation, and the boundless pursuit of exploration.

Underwater and Underground Projects: The Depths of Secrecy

In the clandestine realm of technological innovation and military strategy, underwater and underground projects have long been shrouded in mystery. These secret ventures plunge into the depths of secrecy, where advanced research and development takes place beneath the surface, hidden from public view. Underwater projects, ranging from submarine technology to secret naval installations, have played a vital role in the evolution of maritime warfare and global power dynamics. The strategic importance of underwater warfare has been evident throughout history, influencing naval tactics and geopolitical manoeuvring.

Underground projects have ventured into clandestine construction and infrastructure development, allowing hidden facilities and tunnels to serve various purposes with unparalleled discretion. This subterranean realm offers opportunities for covert operations, secure storage, and concealed transportation networks, shaping the landscape of modern conflict and government operations.

Furthermore, currents of collaboration emerge as nations align or clash in these secret activities, fostering international alliances or covert rivalries. The intertwining of technological advances and geopolitical ambitions further highlights the complex nature of submarine and underground projects, unveiling a world that escapes the common gaze. By unveiling the intrigue surrounding these veiled innovations, we gain a better understanding of the clandestine forces that have shaped our past and continue to shape our future, inviting readers to immerse themselves in the enigmatic world of deep-sea exploration and underground infrastructure.

The underground currents of collaboration: Allies and adversaries

The hidden world of clandestine technological developments often involves a complex network of collaboration and competition, where allies and adversaries dance a delicate and complex tango in pursuit of technological supremacy. Throughout the clandestine projects of the mid-20th century, alliances were formed and shattered, secrets were shared and stolen, and the quest for technological superiority blurred the lines between friends and foes.

At the heart of these underground currents of collaboration lies the multi-level exchange of knowledge, resources, and expertise between nations with aligned interests. While public opinion was focused on geopolitical tensions and ideological divides, behind the scenes, unexpected partnerships emerged, creating a labyrinth of intertwined interests. The gravity of the Cold War fostered unprecedented cooperation between former wartime adversaries, as the race for technological dominance overshadowed historical animosities. At the same time, the spirit of competition that drove technological advances also sowed the seeds of mistrust and subterfuge. Allies turned rivals engaged in a shadow dance of espionage, seeking to gain the upper hand by uncovering each other's hidden innovations. Industrial espionage, infiltration, and sabotage became dangerous tools in the arsenal of allies and adversaries alike, amplifying the clandestine nature of technological advances.

The complex landscape of collaboration was not limited to nations alone. In the closed circles of research and development, individuals, scientists and engineers from diverse backgrounds rubbed shoulders, driven by a common pursuit of cutting-edge technologies. Yet beneath the veneer of collaboration, conflicts simmered, as personal ambitions, loyalty dilemmas and conflicting agendas added layers of complexity to the already intricate tapestry of secret projects. Thus, unravelling the web of allegiances and antagonisms will shed light on the intertwined paradigms of collaboration, competition, and clandestine ambition that defined the discreet realm of hidden technologies.

Speculating on the future: The anticipated impact of hidden technologies

As we delve into the realm of classified technological advances and innovations, it becomes imperative to consider the potential ramifications and future implications of these secret developments. The impact of hidden technologies on global dynamics, security, the economy, and the very fabric of human existence cannot be overlooked. Speculation about the unforeseen consequences of these clandestine projects can offer valuable insights into the trajectory of our collective future. The influence of advanced weapons systems on geopolitical landscapes and strategic balances is one of the key areas that warrants closer examination. The proliferation of secret military technologies can simultaneously trigger arms races, reshape traditional alliances, and spark new forms of global conflict.

The convergence of biotechnological efforts and state secrecy raises ethical concerns and potential risks regarding genetic engineering, biological warfare, and the manipulation of natural ecosystems. The societal, environmental, and geopolitical implications of these hidden biotechnologies require careful consideration. Clandestine advances in communications technologies could redefine the boundaries of privacy, freedom, and surveillance to an unprecedented degree. The fusion of encryption techniques, artificial intelligence, and cyber capabilities could revolutionise the nature of information warfare and influence the dynamics of national and international governance.

Clandestine energy projects shrouded in secrecy are likely to disrupt global energy markets, environmental sustainability, and climate change mitigation efforts. The paradigm shift resulting from the use of undisclosed energy sources has the potential to redefine the global energy landscape and cause unforeseen ecological and economic disruptions. Aerospace innovations hidden from public view promise to reshape the future of aviation, space exploration, and strategic defence capabilities. The emergence of hypersonic flight technologies, advanced propulsion systems, and next-generation spacecraft could redefine humanity's reach beyond the confines of our planet and redefine the boundaries of modern warfare.

Submerged and underground projects shrouded in secrecy raise pressing questions about maritime security, resource exploitation, and territorial claims. The development of stealth underwater vehicles, undetectable submersible platforms, and concealed underground facilities poses new challenges for maritime domain awareness, strategic deterrence, and the enforcement of international maritime law.

In short, the anticipated impact of hidden technologies extends far beyond the veil of secrecy. It warrants thorough exploration, critical analysis, and proactive discourse to anticipate and ethically navigate the complexities of an increasingly clandestine technological landscape.

14
The Industrial Powerhouse
Techniques and transformation

Industrial Progress

The role of technological advances in developing industrial capabilities during wartime cannot be underestimated. The evolution of industrial warfare necessitated a radical transformation of manufacturing processes, production capabilities, and logistical infrastructure. Here, we examine the critical role played by innovative technologies in strengthening industrial prowess during times of conflict.

The period leading up to and following the Second World War marked a turning point in industrial progress. As the demands of war escalated, traditional production methods and capacities became inadequate. It was therefore imperative to adopt new technologies that could streamline manufacturing, improve efficiency and increase output. From the implementation of assembly line techniques pioneered by Henry Ford to the integration of mass production methodologies, these advances revolutionised industrial capabilities on a global scale. The rapid development of highly specialised tools, machinery and equipment also contributed significantly to the expansion and modernisation of industrial infrastructure. Technological innovations such as the introduction of automation, the incorporation of precision engineering, and the use of advanced materials have led to remarkable improvements in overall productivity and production quality.

The convergence of engineering, science, and industry catalysed the creation of advanced weapons, vehicles, and machinery that played a decisive role in achieving strategic superiority on the battlefield. The symbiotic relationship between technological progress and industrial transformation has undeniably played a central role in fuelling the war machine, with nations vying to outdo each other in terms of production and innovation. Furthermore, collaborative efforts between government agencies, research institutes and private companies have facilitated the rapid deployment and scaling up of essential advances. This orchestrated approach accelerated the pace of industrial adaptation and paved the way for sustained growth and development in the aftermath of the conflict. Upon further examination, it becomes evident that the interaction between technology and industry had profound implications not only for the outcome of the war effort but also for the subsequent trajectory of global industrialisation.

Infrastructure and War Production

During the tumultuous period of the Second World War, the industrial landscape underwent a radical transformation in response to the demands of war production. The scale of the conflict necessitated a significant expansion and reconfiguration of existing infrastructure to meet the growing demand for munitions, equipment, and supplies. The war effort catalysed an unprecedented mobilisation of resources and expertise, forcing nations to rapidly adapt their production capabilities to support the impending global conflagration.

To meet the pressing needs of war production, governments and private industries mobilised resources to build new factories, expand existing facilities, and retool assembly lines. Massive investments were made to modernise production sites, integrate cutting-edge technologies, and streamline processes for maximum efficiency. The strategic location of these facilities played a critical role in protecting supply chains and mitigating vulnerabilities to enemy attacks, while ensuring the timely delivery of essential equipment to the front lines. The scale and scope of war production required close collaboration between the public and private sectors, resulting in innovative partnerships that blurred the conventional boundaries between industry and government. The integration of scientific research, engineering prowess, and industrial expertise fostered revolutionary advances in metallurgy, chemistry, and manufacturing techniques, all geared toward achieving superior product quality and quantity.

The imperatives of wartime production stimulated the development of new methodologies such as assembly line production, just-in-time logistics and mass customisation, paving the way for the post-war industrial revolutions. Furthermore, the intensification of war production posed formidable challenges in terms of labour, with vast waves of men and women being drawn into the workforce, often replacing those who had enlisted in the armed forces. This influx strained traditional labour practices, requiring innovative solutions to ensure sustained productivity while preserving worker well-being. Technological innovations, including the implementation of automated machinery and mechanised processes, revolutionised the nature of industrial labour, creating lasting ramifications for the global workforce.

In sum, the era of wartime production witnessed a radical transformation of manufacturing infrastructure and practices, redefining the contours of industrial prowess. Seismic shifts in resource utilisation, technological convergence, and labour dynamics sustained the war effort and shaped the trajectory of industrial evolution for decades to come, leaving an indelible legacy on the fabric of modern industrial civilisation.

Revolution in manufacturing processes

The period surrounding the Second World War saw a monumental transformation in manufacturing processes, marking a turning point in industrial history. Faced with increased demand for military equipment and goods, traditional manufacturing methods proved inadequate to meet the growing needs of war production. This led to a paradigm shift in the way goods were produced, laying the foundations for modern manufacturing techniques that continue to shape industries today. Central to this revolution was the introduction of mass production systems, pioneered by visionaries such as Henry Ford and perfected by industrial leaders eager to streamline production lines. The assembly line, once confined to automobile manufacturing, became emblematic of the era, enabling factories to produce goods at an unprecedented rate. By breaking down complex tasks into simpler, repetitive operations, manufacturers achieved significant gains in productivity and efficiency, ensuring a steady supply of essential wartime supplies.

Advances in science and materials engineering ushered in a new era of innovation. Metals, plastics, and composite materials underwent rigorous research and development, resulting in the creation of durable, lightweight components essential to military vehicles, aircraft, and weaponry. These breakthroughs not only transformed the composition of goods, but also improved their performance and durability on the battlefield. Alongside these advances, the proliferation of machine tools and automation technologies has fostered the evolution of precision manufacturing. Computer numerical control (CNC) machines, for example, have enabled the automated manufacture of complex parts with unparalleled precision, marking a break with traditional manual labour.

The adoption of statistical quality control methods, championed by figures such as W. Edwards Deming, improved the standardisation and consistency of manufactured products, fostering a culture of continuous improvement and excellence in production processes. This era was also marked by the convergence of interdisciplinary knowledge, with scientific discoveries merging with industrial applications. Fundamental research in fields such as chemistry, physics and materials science found practical resonance in manufacturing, fostering the creation of entirely new product categories. From synthetic rubber and high-octane fuels to specialised alloys and electronic components, the war effort catalysed an unprecedented surge of technological ingenuity, reshaping the landscape of consumer and industrial goods.

In short, the revolution in manufacturing processes during this pivotal period produced lasting legacies that extend far beyond the war years. It laid the foundations for modern industrial practices, heralding an era characterised by mechanised precision, standardised quality and the relentless pursuit of innovation. The impact of these transformative changes reverberates across a myriad of sectors, summarising the indelible mark left by wartime industrial prowess on the tapestry of human progress.

Resource Management Strategies

During the era of wartime industrial production, effective resource management strategies played a critical role in maintaining the massive output required to support the war effort. That is why we are examining the complex web of logistical challenges and strategic decisions that shaped the allocation and use of resources during this critical period. One of the fundamental aspects of resource management was the judicious allocation of raw materials, including steel, aluminium, and other essential metals. The acquisition, distribution, and allocation of these resources were governed by meticulous planning aimed at optimising their use across various production sites.

Managing labour resources was a complex task that involved mobilising the workforce, developing skills and ensuring the well-being of workers. Labour efficiency and productivity were essential to meet the increased demands of war production. Furthermore, the management of energy resources such as coal, oil and electricity required innovative approaches to minimise waste and maximise production. Various initiatives were taken to rationalise energy consumption and explore alternative sources to meet the industry's appetite for energy. Another crucial element was transport logistics, which encompassed the movement of raw materials, components and finished products between dispersed production facilities. Optimising transport networks helped to avoid bottlenecks and ensure a steady flow of resources.

Furthermore, the maintenance and servicing of machinery and equipment was a key aspect of resource management, with rigorous maintenance schedules and repair protocols implemented to maintain production capabilities. It is important to note that strategic resource management not only facilitated uninterrupted production, but also laid the groundwork for post-war reconstruction and economic recovery. The lessons learned from effective resource management during wartime became invaluable for subsequent industrial transformations, serving as a beacon for future efforts in resource optimisation and sustainable production practices.

Technological synergies and innovations

The interconnection of technological innovations during the war period led to a myriad of advances that reshaped industries beyond the battlefield. Technological synergies were at the heart of this transformation, as various fields such as aviation, engineering, and materials science converged to create revolutionary developments. The fusion of technologies not only revolutionised warfare, it also catalysed long-term changes in civilian sectors. One of the most notable examples of technological synergy is the collaboration between aerospace engineering and electronics, which led to the development of sophisticated radar systems and airborne detection equipment. These innovations not only gave the Allied forces a strategic advantage, but also laid the foundation for modern air traffic control and weather forecasting systems. The interdisciplinary approach to problem-solving fostered unprecedented breakthroughs that transcended the boundaries of traditional expertise.

Materials science also underwent a paradigm shift during this period, with metallurgy and chemistry converging to produce advanced alloys and synthetic materials of unparalleled resilience and versatility. These new materials found applications in aeronautical construction, weaponry, and medical equipment, setting new standards for durability and performance. The cross-pollination of knowledge and techniques from disparate fields propelled industrial capabilities to unprecedented levels, redefining the limits of what was achievable.

The convergence of chemical engineering and pharmaceutical research enabled the synthesis of vital drugs, such as penicillin, and the mass production of therapeutic compounds. The successful integration of these disciplines not only met critical medical needs during wartime, but also paved the way for significant advances in the pharmaceutical industry, illustrating the lasting impact of technological synergies on public health and well-being.

Furthermore, the intersection of engineering and computer science gave rise to early computing devices, such as Turing's bomb, which marked the first steps of modern computing. This collaboration revolutionised data processing and cryptography, laying the foundations for the digital age and influencing the trajectory of information technology. The profound ripple effects of interdisciplinary innovation highlighted the transformative potential of technological synergies in an era marked by upheaval and uncertainty.

In short, the harmonisation of diverse disciplines ushered in an era of unprecedented technological progress, paving the way for multifaceted innovations with far-reaching implications. The legacy of these synergistic efforts testifies to the enduring power of collaboration across different fields of expertise and highlights the limitless possibilities that arise when knowledge converges for the greater good.

Workforce Mobilisation and Challenges

Mobilising the workforce during wartime presented a myriad of challenges and complexities that required innovative solutions. With the demand for skilled labour skyrocketing, industries faced the daunting task of rapidly increasing their workforce while maintaining efficiency and productivity. This necessitated a re-evaluation of traditional recruitment methods and the implementation of large-scale training programmes to equip workers with the necessary skills. The need for speed and scale in workforce expansion created significant logistical and administrative hurdles, forcing organisations to streamline their processes and adopt new approaches to human resource management.

The entry of women into the labour market in roles traditionally held by men led to a transformation of societal norms and work dynamics. The challenges posed by the integration of a diverse and expanded workforce, combined with the unprecedented demands of war production, forced industries to adapt and evolve their organisational structures. Amidst these transformations, the importance of maintaining morale, ensuring fair labour practices, and fostering a sense of unity and purpose among workers became paramount. These efforts were essential to maintaining productivity and strengthening the resilience of the workforce in the face of arduous conditions and intense pressure.

The widespread impact of the war on communities and families highlighted the need for comprehensive support systems and social welfare initiatives to alleviate the hardships faced by workers and their loved ones. Constraints on infrastructure, transport, and housing posed additional obstacles to effective workforce mobilisation, prompting public and private entities to collaborate in addressing these challenges. Despite the formidable obstacles encountered, the era of wartime labour mobilisation served as a catalyst for pioneering advances in human resource management, labour relations and social equality. The lessons learned from this period continue to resonate in modern organisational practices and provide insight into the adaptability and resilience of the workforce in extraordinary circumstances.

Economic impacts and growth

After the cessation of hostilities, the industrial power forged during the war years began to undergo a profound transformation. The economic consequences of this transformation were manifold, reshaping global markets and paving the way for unprecedented growth and development. As nations shifted from wartime production to peacetime activities, industries had to adapt their capabilities and reorient their strategies. This era was marked by the conversion of massive war production facilities, which presented both opportunities and obstacles. One notable effect was an increase in employment, as soldiers returned home and sought to re-enter the civilian workforce.

The shift from military to civilian production opened new avenues for innovation and specialisation, fuelling economic growth through diversification. Technologies developed and refined during the war found applications in various industries, propelling sectors such as aviation, telecommunications and manufacturing into a period of unprecedented expansion.

The post-war years also saw the emergence of new consumer markets, with pent-up demand and increased disposable incomes stimulating consumption. This spurred investment and the need to increase production capacity to meet the growing demand of a burgeoning population. Industrial giants, which had honed their expertise during the war through large-scale contracts, now ventured into new ventures, capitalising on their accumulated knowledge to drive economic progress.

The global economic landscape was significantly altered at the end of the war, with strategic trade partnerships shaping international commerce. Reconstruction efforts in war-torn regions led to significant investment, resulting in a redistribution of wealth and resources on a scale never seen before. While the war took a heavy toll on infrastructure, it also necessitated large-scale reconstruction, which benefited industries specialising in construction, raw materials and engineering services. The effects of this economic boom extended far beyond the industries directly involved in the war effort, fostering a climate of innovation and competition that laid the groundwork for transformative advances in technology and manufacturing processes. The interaction between technological progress and economic growth became a defining feature of the post-war period, ushering in an era of unprecedented prosperity and setting the stage for the modern industrial landscape.

The transition from war to peace

When the dust of war settled, the industrial power that had fuelled the conflict found itself at a critical crossroads. The transition from a war economy to a peacetime economy presents many challenges and opportunities. The sudden shift in demand and priorities required a fundamental restructuring of industries that had been galvanised by the war effort. The industrial complex, once driven by the urgency of war, now had to redefine its purpose and activities in a world seeking stability and reconstruction. A key aspect of this transition was the demobilisation of wartime production and the reintegration of labour and resources into civilian sectors. This involved streamlining manufacturing processes, reallocating facilities, and realigning supply chains to meet the needs of peacetime markets.

The workforce that had been mobilised for the war had to be retrained for civilian roles, which presented a complex set of social and economic challenges. The conversion of technologies and expertise developed during the war to peaceful applications further accentuated the change. Innovations in areas such as aviation, materials science, and communications, which had advanced military progress, now promised to improve productivity, transportation, and infrastructure development in the post-war period. This process of technology transfer and adaptation opened a new chapter in industrial evolution, where lessons learned from the war effort paved the way for unprecedented advances in various civilian fields. Simultaneously, the global landscape of trade and commerce was transformed.

New international relationships emerged, reshaping economic dynamics across borders. The recovery and redistribution of resources previously allocated to the war effort required diplomatic negotiations and engagement in cooperative projects aimed at promoting mutual benefits and stability. Reconstruction and rehabilitation efforts demanded unprecedented collaboration and innovation, fostering new paradigms of economic interdependence and development. In essence, the transition from war to peace marked a turning point in history, where industrial prowess and capabilities were reshaped to serve the greater cause of prosperity and progress. This metamorphosis involved not only the adjustment of material goods and infrastructure, but also a collective recalibration of society's values and aspirations. The lessons learned from this period of transition continue to offer illuminating narratives and principles for managing industrial transformations in today's dynamic global landscape.

Conclusion: Lessons for the Industry of the Future

When we reflect on the profound transformations that occurred during the transition from war to peace, it becomes clear that valuable lessons can be learned for the future of industry. The period following wartime production was marked by a rapid shift in priorities, as nations around the world refocused their efforts on reconstruction and the establishment of sustainable economic systems. This phase laid bare the resilience and adaptability of industrial sectors, offering invaluable insights into the potential for transformative change. The lessons learned from this transitional period have lasting significance for the trajectory of future industries. One of the key aspects of this transition is the need for diversified innovation. The war required an unprecedented mobilisation of resources and talent, which led to revolutionary advances in various fields. As industry shifted to peacetime, the challenge was to harness this momentum and channel it into diverse areas such as infrastructure, technology, and consumer goods. This diversification highlighted industry's ability to pivot quickly and respond to changing demand, a phenomenon that has direct implications for future industrial landscapes.

Furthermore, the transition from war to peace highlighted the imperative of sustainable resource management. During wartime production, resource conservation and optimisation emerged as a key priority, supporting military efforts. However, when the focus shifted to reconstruction and growth, the resource management paradigm underwent a substantial transformation. The principles of efficiency and sustainability that were refined in the crucible of wartime demands continue to resonate as guiding principles for future industries, underscoring the enduring importance of responsible resource use. The socio-political dynamics that developed during the transition from war to peace also provided relevant lessons for the industry of the future. The reintegration of the workforce, along with the restoration and reconciliation of society, was a multifaceted undertaking that highlighted the symbiotic relationship between industrial progress and overarching social goals. The strategies and policies adopted at that time not only facilitated economic rejuvenation, but also set precedents for inclusive growth and equitable opportunities in industrial settings. These foundations are reflected in the future landscape of industries, underscoring the fundamental role of human-centred approaches and the promotion of resilient and adaptive communities.

In short, the transition from war to peace demonstrates the remarkable capacity of industries to evolve, innovate and bring about significant change on a global scale. The lessons learned from this phase of transformation serve as a guide for the future trajectory of industries – a legacy imbued with vital ideas, heralding the promise of dynamic, sustainable and inclusive industrial practices for generations to come.

15
Spy Games

Espionage during the Second World War

Espionage during the Second World War represents a complex and intriguing facet of the global conflict, characterised by clandestine operations and intelligence gathering with considerable stakes. The origins of espionage can be traced back to the urgency and necessity for nations to gain a strategic advantage over their adversaries. In the crucible of war, the art of espionage emerged as an essential tool for understanding the enemy's intentions, military capabilities and political manoeuvres. The motivations for espionage activities during this period were deeply rooted in the imperatives of protecting national interests, ensuring military superiority, and obtaining vital intelligence to inform strategic decisions. As countries found themselves embroiled in a devastating global conflict, the need for accurate and timely information became paramount, leading to an unprecedented expansion of espionage networks and a refinement of the art of espionage. The immense pressure to gather actionable intelligence, penetrate enemy lines, and decode encrypted messages spurred the development of sophisticated spy networks and secret agents.

The moral and ethical dimensions of espionage were uniquely tested during the Second World War, as agents grappled with the complexities of espionage, including the moral dilemmas inherent in deception, sabotage, and subterfuge. This chapter aims to delve into the fascinating history of espionage during the Second World War, highlighting the multifaceted motivations, operational dynamics, and lasting impact of clandestine intelligence activities that reshaped the course of history.

Recruitment and Training of Spies

The recruitment and training of spies during the Second World War was a crucial and secretive process that played a vital role in the unfolding of history. Identifying potential agents with the necessary skills and attributes required meticulous attention to detail, as well as a thorough understanding of the geopolitical landscape. Intelligence services scoured various sectors of society to identify individuals who possessed the qualities of discretion, ingenuity, and unwavering commitment to their mission. Recruits often came from diverse backgrounds, including academia, business, and the military, allowing for a wide range of expertise and skills to be leveraged. Once identified, potential agents undergo a rigorous selection process, during which their ability to maintain secrecy and cope with intense pressure is thoroughly assessed. Those who demonstrated exceptional potential were then subjected to intensive training programmes specifically designed to cultivate their espionage skills. These programmes covered a wide range of topics such as surveillance techniques, cryptography, operational security and psychological manipulation.

In addition to technical skills, recruits were also trained in the art of blending seamlessly into different cultures and circumstances, and often took language immersion and cultural awareness courses to ensure their effectiveness in foreign environments. They also learned the intricacies of creating and maintaining cover, establishing secure channels of communication, and avoiding detection by enemy counter-intelligence services. The psychological preparation of spies was equally crucial, with trainees receiving training in the mental fortitude necessary to withstand interrogation, endure isolation, and make life-and-death decisions in a split second. Once properly trained, these agents were strategically deployed across enemy lines and territories, where they carried out missions crucial to the war effort. The recruitment and training of spies remains a fascinating testament to the clandestine nature of espionage during the Second World War, highlighting the meticulous preparation and extraordinary courage that characterise these unsung heroes of history.

Tools of the trade: gadgets and technology

In the world of espionage during the Second World War, technological advances played a vital role in providing spies with tools that were not only innovative, but often ahead of their time. These gadgets enabled clandestine agents to gather intelligence, conduct covert operations, and communicate securely, ultimately influencing the course of the war. One of the most iconic devices used by spies was the Enigma machine, a complex encryption tool used by the Axis powers to code sensitive messages. The decryption efforts of code breakers, including Alan Turing, revealed enemy plans and gained strategic advantages. In addition, advanced cameras concealed in everyday objects, such as cigarette cases and buttons, allowed spies to discreetly take photographs of classified documents and facilities. These miniature but powerful cameras revolutionised the field of intelligence gathering, enabling covert reconnaissance and the acquisition of vital information.

Furthermore, disguised weapons, ranging from pen guns to lipstick guns, provided agents with clandestine means of self-defence during high-stakes encounters. Radio transmitters disguised as innocuous objects provided crucial communication channels for undercover agents, allowing them to transmit sensitive data without arousing suspicion. The development of microfilm technology also facilitated the concealment and transport of large volumes of documents, ensuring the secure transport of critical intelligence. The complexity and ingenuity of these gadgets underscore how deeply technology became intertwined with the art of espionage, illustrating the ingenuity and innovation displayed by both Allied and Axis intelligence services. As we delve deeper into the field of espionage technology, it becomes clear that these advances reshaped the landscape of wartime espionage, reinforcing the importance of pioneering inventions in shaping the outcome of some of history's most significant conflicts.

Strategic Operations: Behind Enemy Lines

During the Second World War, strategic operations behind enemy lines played a vital role in the course of the war. These covert missions were meticulously planned and executed to disrupt enemy activities, gather intelligence, and support resistance movements. The success of these operations often depended on the agents' ability to blend in with the local population and operate covertly in hostile territory. Behind enemy lines, agents faced many challenges, including navigating unfamiliar terrain, evading enemy patrols, and communicating securely with their superiors. They relied on their ingenuity, adaptability, and extensive training to carry out their missions while minimising the risk of detection. To achieve their objectives, operatives often used a combination of techniques, such as disguise, sabotage, and clandestine communications.

One of the main strategic operations carried out behind enemy lines was to sabotage the enemy's infrastructure and supply lines. Specialised teams were tasked with disrupting enemy transport networks, destroying key facilities and sabotaging vital supply routes. These acts of sabotage not only inflict direct damage on the enemy, but also sow confusion and disarray in their ranks, thereby contributing to the overall war effort. In addition to sabotage, intelligence gathering is another essential aspect of strategic operations behind enemy lines. Operatives are trained to observe and report on enemy movements, gather information on enemy deployments, and assess the morale and capabilities of enemy forces. This valuable intelligence provided Allied commanders with vital information that influenced their tactical and strategic decisions.

Behind enemy lines, agents worked closely with local resistance groups, forging partnerships and providing support to indigenous forces. Collaboration with local networks allowed agents to leverage local knowledge and establish channels for disseminating information and coordinating actions. This collaboration fostered a sense of solidarity and inspired resilience among local populations, strengthening their determination to resist enemy occupation. The successful execution of strategic operations behind enemy lines required precision, discretion, and unwavering commitment. These secret missions, often carried out under cover of darkness and in the face of grave danger, illustrate the bravery and ingenuity of the agents who dared to venture into hostile territory to wage a different kind of battle—one fought in the shadows, where victory depended on secrecy, cunning, and courage.

Key figures in spy networks

During the Second World War, the success of espionage and intelligence operations depended on the dedication and skill of key figures in spy networks. These individuals often operated in the shadows, demonstrating remarkable courage and ingenuity in their quest for crucial information. One such figure was Virginia Hall, an American spy who overcame numerous obstacles, including a prosthetic leg, to become one of the most effective Allied agents in occupied France. Her efforts helped support the resistance movement and gather vital intelligence for the Allies. Juan Pujol Garcia, a Spanish double agent who played a key role in deceiving the German high command about the location of the D-Day invasion, is another notable figure. Working under the code name 'Garbo,' his ability to fabricate convincing misinformation saved countless lives and ensured the success of the Allied operation. Spy networks also benefited from the remarkable talents of Noor Inayat Khan, a British Muslim who showed extraordinary bravery as a radio operator in

Nazi-occupied France. Despite the danger, she continued her clandestine communications until she was captured by the Gestapo, enduring harsh interrogation without revealing any crucial information. These are just a few examples of the remarkable individuals whose contributions to espionage networks changed the course of history during this tumultuous period. Their stories are testament to the selflessness and ingenuity that characterised espionage efforts during the Second World War.

Communication Methods and Code Breaking

During the Second World War, effective communication methods and code breaking played a crucial role in the success of espionage and military operations. Both the Axis and the Allies used complex codes and encryption systems to transmit vital information while attempting to intercept and decipher their opponents' messages. One of the most famous examples of successful code-breaking is that of the British team at Bletchley Park, which cracked the code of the German Enigma machine, providing the Allies with invaluable intelligence. The use of encryption devices such as Enigma and Lorenz SZ40/42 by the Axis powers required innovative approaches to intercept, decipher, and exploit their coded communications. Conversely, clandestine agents used various forms of hidden messages, including invisible inks, microdots, and even seemingly innocuous everyday objects containing secret information. These secret communication methods required ingenuity and resourcefulness on both sides.

The evolution of cryptography and cryptanalysis during this period marked a turning point in the history of information security and intelligence gathering, laying the foundation for modern cryptography and communications security. The resources invested in decrypting and intercepting communications illustrate the importance of confidentiality and the efforts made by nations to safeguard and obtain critical information. By understanding the intricacies of communication methods and codes, we gain a better understanding of the intelligence war that shaped the outcome of World War II and its impact on modern cybersecurity and espionage tactics.

Infiltration and counter-espionage techniques

Infiltration and counter-espionage techniques played a vital role during the Second World War, influencing the course of history through clandestine operations and strategic espionage. Infiltration involved the discreet insertion of specially trained agents into enemy territory for the purpose of gathering vital intelligence, disrupting enemy activities, and carrying out covert missions. The success of these operations depended largely on the ability to remain undetected while blending seamlessly into the local environment.

Counter-espionage, on the other hand, encompasses defensive measures designed to thwart enemy infiltration and protect classified information. It involves identifying and neutralising enemy spies and implementing security protocols to protect sensitive military, political and technological resources. Counter-espionage efforts have also focused on detecting and decrypting enemy codes and encryption algorithms to prevent unauthorised access to crucial communications. Infiltration techniques varied widely, ranging from the use of undercover agents posing as civilians or refugees to the insertion of highly trained special forces behind enemy lines. Agents used false documents, disguises, and language skills to blend in with the local population while conducting reconnaissance operations and transmitting crucial information to their superiors. The ability to maintain cover under scrutiny and interrogation is essential to the success of these agents, who often must demonstrate nerves of steel and quick thinking in unpredictable situations.

Successful infiltration required meticulous planning and coordination, as well as detailed knowledge of the target area, its inhabitants, and the enemy's modus operandi. This in-depth knowledge enabled agents to navigate complex terrain, establish safe houses, and gather intelligence without arousing suspicion. In some cases, agents had to endure long periods of isolation and extreme danger, demonstrating remarkable resilience and commitment to their mission despite insurmountable obstacles. Counter-espionage strategies relied on a multi-faceted approach, including surveillance, analysis of intercepted enemy communications, and the development of complex deception operations designed to mislead enemy agents. The use of double agents and the dissemination of false information were tactics frequently employed to deceive enemy intelligence services and disrupt their activities.

Strict security measures were implemented to protect key military installations, research centres, and command centres from infiltration or sabotage by enemy agents. The interplay between infiltration techniques and counter-espionage has shaped the delicate balance of power in wartime, influencing the outcome of decisive battles and the overall trajectory of conflict. Intelligence gleaned from successful infiltrations has provided crucial information that has guided military strategies, targeted high-value enemy assets, and bolstered the morale of allied forces. Conversely, effective counter-espionage efforts prevented vital information from falling into enemy hands, preserving strategic advantage and mitigating the impact of espionage on the war effort. The complex dance of infiltration and counter-espionage illustrates the high stakes of espionage, where every move has profound repercussions on the outcome of the war. Throughout the conflict, innovations in offensive and defensive espionage strategies continued to evolve, shaping the changing landscape of clandestine operations and leaving an indelible mark on the annals of history.

Major Espionage Cases: Successes and Failures

Throughout World War II, espionage played a critical role in the outcome of crucial battles and strategic decisions. That is why we pause to examine some of the most important espionage cases that influenced the course of the war, highlighting notable successes and failures.

One such case is Operation Mincemeat, a highly successful British deception strategy that misled the Axis powers about the Allies' invasion of Southern Europe. By placing fake documents on a corpse off the coast of Spain, the Allies successfully deceived German intelligence, diverting Nazi forces from their true target, Sicily. This ingenious operation is celebrated as a classic example of successful espionage, which significantly altered the course of the war. Another notable success was the breaking of the German Enigma code by Allied cryptanalysts at Bletchley Park. This breakthrough provided invaluable intelligence to the Allies, allowing them to anticipate and counter German military movements, which ultimately tipped the balance in their favour.

However, there were also significant failures in espionage that had dire consequences. One infamous case was the capture and subsequent conversion of several double agents by the Soviets within the British and American espionage networks. This betrayal exposed many vital operations and compromised the integrity of Western intelligence services. Furthermore, the failure of Operation Tiger, a planned rehearsal for the D-Day landings, due to a German attack, resulted in significant human casualties and the loss of a crucial secret to the enemy. These failures highlight the high stakes and inherent risks of espionage, and underscore the complex interplay between success and failure in the murky world of clandestine operations. In examining these examples, it becomes clear that espionage during the Second World War was characterised by a precarious balance between triumphs and failures, underscoring the profound impact of secret intelligence on the outcome of historical events.

Impact of Espionage on the Outcome of the War

Espionage played a decisive role in the outcome of the Second World War. The collection and transmission of crucial intelligence significantly influenced strategic decision-making and military operations, ultimately impacting the overall course of the war. By acquiring valuable information about enemy movements, capabilities, and intentions, espionage enabled Allied forces to anticipate and counter Axis offensives. Conversely, Axis intelligence operations sought to undermine Allied efforts and gain superiority through deception and sabotage.

Successful infiltration and reconnaissance missions carried out by skilled agents provided essential information about enemy plans, fortifications, and vulnerabilities, allowing commanders to devise effective tactics and adjust combat strategies in real time. The interception and decryption of encrypted communications, such as the breaking of Enigma codes, revealed vital enemy directives and facilitated pre-emptive actions that often proved decisive in crucial engagements.

The dissemination of false information and disinformation by both sides was intended to confuse and mislead opponents, divert resources, and sow discord among opposing forces. Covert operations also played a role in influencing public opinion, both nationally and internationally, by shaping perceptions of the war effort and garnering support for propaganda campaigns. The impact of espionage extended beyond the battlefield, permeating diplomatic negotiations and shaping post-war geopolitical agreements.

As the war drew to a close, intelligence gathered through espionage guided the planning of final offensives and contributed to the rapid conclusion of hostilities. Knowledge gained through spy networks and reconnaissance missions identified strategic targets and enabled the devastating blows that led to the defeat of the Axis powers. The legacy of wartime espionage reverberated at the site in the post-war period, laying the groundwork for the evolution of intelligence agencies and the formulation of new security doctrines to address emerging global threats. The profound influence of espionage on the outcome of the Second World War confirms its status as an indispensable element of modern warfare, demonstrating the enduring importance of clandestine operations in shaping historical events and preserving national security.

Transition from wartime espionage to Cold War tactics

At the end of the Second World War, the intelligence landscape underwent a significant change as the world entered the turbulent era of the Cold War. The transition from wartime espionage to Cold War tactics marked a new chapter in global espionage, characterised by ever-evolving strategies, heightened tensions, and a complex network of surveillance and covert operations. As former allies became rivals, the Allied powers and their former adversaries were quick to adapt their intelligence-gathering methods to the demands of this new geopolitical climate.

The Cold War saw the emergence of a myriad of clandestine activities fuelled by ideological competition and the pursuit of strategic advantages. Traditional espionage tactics such as infiltration, surveillance, and intelligence gathering continued to play a vital role, but new technological advances began to shape the way information was obtained, disseminated, and protected. The use of innovative communication devices, cryptanalysis, and the burgeoning field of signals intelligence became essential to gaining a competitive advantage.

At the forefront of this evolution was the growing role of national intelligence agencies such as the CIA, KGB, MI6, and their counterparts in other countries. The secretive nature of their activities has often blurred the line between espionage, counter-espionage and psychological warfare. Propaganda, disinformation campaigns and covert operations have become essential tools in the silent struggle for dominance.

At the same time, the advent of the nuclear age has added a new level of complexity to intelligence operations. The race for nuclear superiority fuelled a relentless quest for information on weapons development, deployment strategies, and the intentions of rival powers. This led to an unprecedented focus on scientific and technical intelligence, requiring the recruitment of agents specialising in fields such as physics, engineering, and nuclear science.

The dynamic nature of Cold War intelligence operations also gave rise to the phenomenon of double agents, moles, and defectors, whose actions influenced the course of history on numerous occasions. The infamous cases of renegades such as Kim Philby and Aldrich Ames highlighted the vulnerability of even the most sophisticated intelligence apparatus to internal threats. Trust, betrayal, and the delicate dance of deception defined this tumultuous period, where every piece of information could tip the balance of power.

As the Cold War unfolded, the technological arms race intensified, giving rise to surveillance satellites, aerial reconnaissance, and other high-altitude espionage tools. The ability to monitor adversaries' movements and activities from space revolutionised the scope and scale of intelligence gathering, forever changing the dynamics of international relations. In essence, the shift from wartime espionage to Cold War tactics ushered in an era of unprecedented sophistication and intrigue in the field of espionage.

16
Impact On Modern Industry And Technology

Tracing the technological lineage

The aftermath of the Second World War marked a turning point in human history, with unprecedented technological advances emerging from the ashes of conflict. When we examine the legacy of wartime innovation, it becomes clear that the technical advances and production techniques developed during this turbulent period have undeniably shaped the modern industrial landscape. To understand the profound impact of these developments, it is imperative to trace the historical context surrounding their genesis. The demands of war led to a rapid expansion of manufacturing capabilities, propelling industries to astonishing heights of productivity and precision. The quest for superior weaponry gave rise to relentless innovation, fostering an environment where engineering prowess intersected with cutting-edge production methodologies.

This dynamic synergy laid the foundation for a lineage of technological marvels that continue to resonate in contemporary industrial practices. Exploring this genetic lineage of innovation offers a compelling narrative that not only illuminates the path from past to present, but also provides valuable insights into the evolution of industrial paradigms. By dissecting the gradual metamorphosis from wartime requirements to peacetime applications, we can discern the lasting imprint of engineering and manufacturing triumphs that have permeated virtually every facet of the modern world. Thus, as we embark on this exploration, we are poised to uncover the fascinating tapestry of technological lineage, intricately woven through the fabric of time, illustrating the unwavering influence of the crucible of World War II on the realms of industry and technology.

Technical and production advances

In the aftermath of the Second World War, the world experienced a remarkable surge in engineering and production advances that would redefine industries around the globe. The need for mass production during the war catalysed innovations in manufacturing processes and techniques, giving rise to modern industrial engineering. One of the major advances was the implementation of assembly line production, championed by figures such as Henry Ford. This innovation revolutionised the manufacturing landscape by enabling streamlined and efficient production of consumer goods. Industrial engineers and designers played a crucial role in optimising production lines and developing advanced machinery to meet the growing demands of the post-war economy. This era also saw the widespread adoption of new materials and alloys, driven by the urgent needs of the war effort. The development of high-strength steel, aluminium alloys and advanced polymers not only strengthened military capabilities, but also found numerous applications in civilian sectors, from automotive manufacturing to construction.

The introduction of computer-aided design (CAD) and computer-aided manufacturing (CAM) systems represented a monumental leap forward in engineering processes. These technologies facilitated precision engineering, rapid prototyping, and a smooth transition from design to production. On the other hand, the automation of manufacturing processes has maximised efficiency, reduced margins of error and accelerated the time to market for new products. The post-war period also marked the advent of nuclear energy and its integration into electricity generation. The development of nuclear reactors, driven in part by military research, presented a multitude of engineering challenges and opportunities. This cutting-edge technology ultimately shaped the energy sector and influenced engineering practices, paving the way for advances in safety protocols, materials science, and the exploitation of alternative energy sources.

The nascent field of aerospace engineering experienced exponential growth during this period. The expertise gained from wartime aeronautical innovations fuelled rapid advances in commercial and military aerospace technologies. The emergence of supersonic flight, composite materials, and fly-by-wire flight control systems revolutionised air travel and redefined the limits of human exploration.

In short, the post-war period witnessed an era of transformative advances in engineering and production techniques that left an indelible mark on virtually every aspect of modern life. The cross-fertilisation of military and civilian technologies paved the way for the contemporary engineering landscape, illustrating humanity's ability to transform the remnants of conflict into catalysts for progress and innovation.

From War to the Workplace: Military Innovations for Civilian Use

Military innovations have always played an important role in the evolution not only of the battlefield, but also of various aspects of civilian life. The transition of military technologies to civilian applications has been a key driver of progress in many industries. From the development of radar and sonar to track enemy movements in wartime to their adaptation for weather forecasting, navigation systems, and marine exploration, the impact of these advances is profound.

One of the most notable examples of this transfer is the integration of GPS technology into everyday devices and applications. Originally developed for military purposes to ensure accurate navigation and tracking, GPS has revolutionised countless industries, from transport and logistics to agriculture and emergency services. Similarly, advances in materials science and manufacturing techniques, initially designed for military hardware, have found widespread civilian application, driving innovation in consumer electronics, automotive engineering, and medical device manufacturing.

Research and development investments by defence agencies have often catalysed breakthroughs in fields such as aerospace, telecommunications, and energy. Take, for example, pioneering work on unmanned aerial vehicles (UAVs) or drones, which have moved beyond intelligence and reconnaissance in military operations to applications in agriculture, infrastructure inspection, and environmental monitoring.

The influence of military innovation extends beyond hardware to encompass software and computing capabilities. Encryption technologies, originally designed to secure military communications, are now integral to protecting sensitive personal and financial information in the digital age. The rapid processing power and data storage requirements for military applications have led to advances in high-performance computing and cloud technologies, with myriad civilian uses ranging from scientific research to entertainment and e-commerce. As we continue to witness the evolution of warfare and defence strategies, there is inherent potential for spillover of advanced technologies into civilian domains. Responsible and ethical considerations surrounding the use of military-derived innovations in non-military spheres remain paramount and require thoughtful dialogue and regulation to maximise benefits to society while mitigating potential risks. Ultimately, the convergence of military and civilian technologies underscores the interconnected nature of progress and the lasting impact of defence innovation on the fabric of modern society.

Electronics and computing: Direct descendants

The impact of the Second World War on modern industry and technology is most clearly illustrated by the evolution of electronics and computing. The urgency of developing advanced communications and information processing systems during wartime catalysed significant advances that have since reshaped the world. The advanced radar and cryptography technologies employed during the war spurred the development of the first computers, laying the foundation for the digital age. ENIAC, one of the first general-purpose electronic digital computers, was a direct product of wartime research and illustrates the vital role that military technological innovation played in the birth of the computing revolution. Beyond hardware, the war also fostered substantial advances in software and algorithm development, as evidenced by the breakthroughs made by decrypters and cryptanalysts. The legacy of these developments lives on today in our dependence on sophisticated computer systems and digital technologies in all sectors of activity.

Efforts to miniaturise and make electronic components more durable during wartime led to the creation of rugged and reliable electronic devices. The emphasis on robustness and efficiency laid the foundation for the consumer electronics industry, whose applications range from household appliances to personal electronic devices. The refinement of vacuum tube technology for military applications not only ushered in the era of mass-produced radios and televisions, but also paved the way for the subsequent transistor revolution that forms the basis of modern computing and telecommunications.

The search for secure and efficient communication systems led to the development of the first networking and data transmission protocols, marking the early stages of what would later become the internet and the global telecommunications infrastructure. In addition to hardware and network infrastructure, the war necessitated accelerated research in materials science and semiconductor physics, which led to fundamental knowledge that propelled the semiconductor industry to its current level. The integrated circuit, a crucial invention born out of this drive for compact and reliable electronic devices, is now the basis for nearly all modern electronic devices, from smartphones to advanced medical equipment. The war's impetus for faster computing and data processing sparked a perpetual cycle of innovation and progress that continues to shape the technological landscape of the 21st century. As we enter the era of artificial intelligence and quantum computing, it is essential to recognise the historical foundations of these advances in the urgent imperatives born out of the crucible of global conflict.

Transformations in the chemical industry: From explosives to pharmaceuticals

The impact of wartime advances in chemistry on modern industry cannot be overstated. It was during the years of conflict that monumental transformations took place in the field of chemical engineering, creating a profound ripple effect that continues to shape many industries today. One of the most significant changes occurred during the transition from wartime explosives production to the pharmaceutical industry. The synthesis of various chemical compounds for use in explosives and chemical warfare agents accelerated the search for new chemical processes and applications.

This surge in chemical innovation paved the way for the development of vital medicines and revolutionary treatments. Advances in synthetic chemistry during the war, which were aimed at harnessing explosive power, laid the foundation for the mass production of antibiotics, hormones, and other pharmaceuticals essential to modern healthcare. Today, many pharmaceutical compounds are direct descendants of chemicals originally designed for the war effort. The mastery of complex chemical reactions and efficient production methods initially developed for warfare have been perfectly adapted to the manufacture of medicines essential for combating disease. The strategic knowledge gained during wartime catalysed the evolution of chemical engineering, enabling the large-scale synthesis and purification of various compounds.

The knowledge gained from studying the toxicological effects of chemicals in military contexts provided critical insights into pharmacology and toxicology. The shift from explosives-focused chemical engineering to pharmaceutical advances represents one of the lasting legacies of wartime innovation. This transformation demonstrates the adaptability and resilience of scientific knowledge, which has moved from a destructive function to one of sustaining life and well-being. The ethical implications of such a transition continue to stimulate discourse on the dual-use potential of technological developments arising from war. Exploring these intersections between war technologies and civilian applications provides insight into the lasting impact of historical events on today's industries.

The Evolution of the Automobile: Impacts of High Speed

In the aftermath of the Second World War, the automotive industry underwent a transformative evolution that reverberated throughout economies and societies around the world. As wartime-proven technology was repurposed for civilian use, demand for high-performance vehicles exploded. The emergence of a dynamic consumer culture further reinforced this paradigm shift, fuelling an unrelenting quest for speed and luxury. This insatiable appetite for innovation and progress drove manufacturers to revolutionise their technical approaches, which had a monumental impact on modern industry and technology. The impact of high speed extended beyond simple performance metrics, encompassing critical aspects such as safety, efficiency, and environmental considerations. The introduction of advanced materials and manufacturing processes enabled the development of lighter and stronger vehicle structures, heralding a new era of crash-resistant designs. At the same time, revolutionary advances in aerodynamics and propulsion systems have enabled unprecedented fuel efficiency, in line with the growing emphasis on sustainable automotive solutions.

The advent of high-performance automobiles has had a profound influence on popular culture, embodying the intersection of technology and lifestyle. Iconic models from renowned manufacturers have become symbols of status and aspiration, transcending their utilitarian function to embody the pinnacle of technical prowess and sophistication. Motorsport events, once reserved for a niche audience of enthusiasts, have become mainstream, captivating audiences with adrenaline-fuelled spectacle and showcasing human ingenuity and ambition.

This transformed landscape paved the way for symbiotic relationships between the automotive industry and ancillary industries, sparking new collaborations and cross-pollination of expertise. In particular, the convergence of automotive and aerospace technologies accelerated the development of cutting-edge features, blurring traditional boundaries and propelling the collective pursuit of excellence. The strategic integration of computer-aided design and advanced materials has given rise to vehicles that have challenged conventional limits, setting new benchmarks in performance, luxury, and safety.

At the dawn of the 21st century, the evolution of the automobile continues unabated, driven by an unwavering commitment to innovation and sustainability. The disruptive forces of electrification and autonomous driving are recalibrating the very essence of automotive transportation, promising a future that is as exhilarating as it is conscientious. The foundations laid during the post-war period have indelibly shaped the trajectory of the automotive industry, imprinting a legacy that resonates with every rev of a powerful engine.

Aerospace Developments: From Rockets to Commercial Jets

The evolution of aerospace technology, from a wartime necessity to a driving force of global connectivity, is a testament to human ingenuity and perseverance in innovation. In the post-war era, advances in rocket and jet propulsion, born out of military research, propelled humanity towards new frontiers in exploration and transport. The development and refinement of rocket technology not only enabled rapid advances in space exploration, but also laid the foundations for revolutionary changes in commercial aviation. Companies that had previously focused solely on the production of warplanes turned their attention to the burgeoning civilian market, giving rise to iconic names in the aerospace industry. With the escalation of the Cold War, the race to conquer the skies spurred unprecedented investment in aerospace engineering. The quest to push the boundaries of speed, altitude and payload capacity paved the way for the genesis of supersonic and, later, hypersonic flight. The transition from military aircraft to commercial jets marked a turning point in global mobility, with a dramatic transformation in passenger transport. The shift to jet airliners revolutionised air travel, making distant destinations more accessible and reshaping international trade and tourism.

Technological advances stemming from the development of missiles and satellites contributed significantly to the accuracy and reliability of navigation systems. The integration of GPS and advanced avionics raised safety standards, reducing the margin for error and improving operational efficiency. At the same time, innovations in materials science, initially intended to improve the performance of military aircraft, found applications in civil aircraft construction, resulting in lighter, stronger and more fuel-efficient aircraft. Ethical considerations also played a key role in the convergence of military aerospace developments with commercial aviation.

As society became more aware of the impact on the environment, the aerospace industry faced increasing pressure to optimise fuel consumption and reduce emissions. This led to a wave of research and design strategies focused on creating environmentally friendly propulsion systems and sustainable aircraft configurations, thus reconciling progress with environmental responsibility. The legacy of aerospace advances extends beyond the realms of innovation and commerce. It has fostered intercultural exchange, broadened horizons, and brought the world closer together. The ethics that once guided the design of warplanes have evolved into a commitment to safe, efficient, and environmentally friendly air travel. The saga of aerospace development, from rockets to commercial jets, testifies to the lasting influence of wartime technologies in shaping the modern world.

Ethical considerations in contemporary applications

The integration of wartime technologies into modern industry and technology has raised complex ethical considerations that require careful thought and examination. As society exploits advances in aerospace, automotive, computing, and other fields influenced by wartime innovations, the ethical implications are considerable. One of the most pressing concerns is the extent to which these technologies are exploited for civilian purposes rather than military or surveillance purposes. The responsible and ethical use of advanced technologies is at the forefront of public discourse and policy-making. In the context of aerospace developments, ethical considerations encompass issues such as the use of drone technology for commercial and military purposes, sparking debates about privacy, security and the risk of misuse.

In the field of computing and artificial intelligence, issues relating to data privacy, algorithmic bias and the societal impact of automation have become paramount. The legacy of wartime chemical and biological research also raises ethical dilemmas when applied to modern pharmaceutical and agricultural sectors, prompting close scrutiny of ecological impacts, health risks, and the prudent management of potentially dangerous substances.

Advances in automotive engineering, particularly the development of high-speed vehicles and autonomous driving technologies, require ethical discussions about safety, liability, and the need for regulations to mitigate potential harm. Furthermore, as innovations from wartime research continue to shape contemporary industry, companies face ethical challenges related to intellectual property rights, corporate responsibility, and accountability in the adoption of technologies. Case studies of companies that have integrated wartime technologies provide a compelling lens through which to examine the ethical dimensions of exploiting historical innovations and the responsibilities associated with doing so. These case studies offer valuable insights into decision-making processes, the balance between societal benefits and risks, and the strategies employed to ensure principled technology adoption. Assessing the ethical implications of integrating wartime innovations into modern industry and technology is crucial to guiding responsible and sustainable progress. As historical precedents show, proactively addressing these ethical considerations can lead to more equitable, safer, and more ethically grounded technological advances that contribute positively to society.

Case studies: Business histories and technology adoption

To study the impact of Second World War technology on modern industry and technology, it is essential to examine specific cases where companies played a key role in the adoption and integration of military innovations into civilian applications. These case studies not only provide insight into technological evolution, but also highlight the organisational and strategic decisions that shaped contemporary industrial landscapes. One such case study is that of IBM, which during the war years was heavily involved in the production of tabulating equipment for the Allies. After the war, IBM successfully transitioned to commercial computing, becoming a pioneer in electronic data processing systems and profoundly influencing the modern computing era. Another fascinating example is the evolution of Volkswagen.

Originally designed by Ferdinand Porsche as a car for the masses under the Nazi regime, Volkswagen's Beetle became an iconic symbol of post-war economic renewal and a classic example of military technology finding civilian use. At the heart of these case studies is the ability of companies to recognise the potential of transformational technologies and exploit them for commercial advantage. A closer look takes us to the aerospace industry, where companies such as Boeing and Lockheed Martin have been pioneers in military and civil aviation. The successful adaptation of rocket propulsion and aerodynamic advances from wartime developments paved the way for commercial jet transport, revolutionising global transportation.

The chemical industry reveals intriguing transformations, with companies such as Bayer leveraging their wartime expertise to develop synthetic materials for use in consumer products and pharmaceuticals. By examining contemporary implications, these case studies highlight the considerable influence of wartime technologies on various industries, illustrating the interaction between historical legacies and today's innovations. Understanding these historical precedents provides insight into the dynamics of technology adoption, competitive strategies, and market positioning, which ultimately shape the trajectory of modern industry and technology.

Conclusion: Ongoing Influence and Future Trajectories

After this exploration of the impact of wartime technologies on modern industry and technology, it is clear that the legacy of conflict has left an indelible mark on the trajectory of innovation. The case studies presented illustrate how companies have adapted, and adopted technologies developed during wartime to revolutionise various sectors, from engineering and manufacturing to healthcare and communications. These adaptations demonstrate the resilience and adaptability of industries that integrate military innovations into products and processes that benefit society as a whole. Looking ahead, it is clear that the influence of war technology will continue to shape the landscape of modern industry and technology. One of the key trajectories lies in the area of ethical considerations. As we move forward, there is an urgent need to address the ethical implications of exploiting technologies derived from warfare. Responsible and conscientious use of these advances will be essential to ensure that potential negative consequences are minimised while maximising the benefits to humanity.

The continuing influence of warfare technology will fuel the ongoing integration of advanced engineering concepts into everyday applications. This integration will catalyse the evolution of technologies across multiple sectors, further blurring the boundaries between traditional silos and fostering interdisciplinary collaboration. The convergence of disciplines will undoubtedly lead to revolutionary innovations that will redefine current paradigms in technology and industry. Specifically, as advances in aerospace continue to permeate the commercial and private sectors, the future trajectory points to unprecedented developments in sustainable air transport, space exploration, and advanced aerial technology. The ongoing influence of warfare technologies is being felt across the economic landscape, contributing to global competitiveness and economic growth. Sectors born out of military innovations serve as catalysts for job creation, investment opportunities, and long-term economic development.

Ultimately, the impact on modern industry and technology is not just a historical analysis, but an ongoing narrative that informs the choices made and directions pursued by individuals and organisations. In essence, the legacy of wartime technology remains a significant force shaping the present and future of our technological landscape.

17
The Legacy of Resistance

Defining resistance

Resistance, in its diverse and multifaceted manifestations, has taken on a variety of interpretations and meanings across different historical and socio-political contexts. The concept of "resistance" transcends simple opposition and encompasses collective and individual acts of defiance, resilience and subversion against oppressive forces, authoritarian regimes and occupying powers.

It embodies the spirit of defiance and covers a spectrum ranging from overt acts of revolt to subtle forms of non-cooperation and civil disobedience. Historical examples of resistance movements have played a decisive role in the outcome of conflicts and wars, imbuing the annals of human history with tales of bravery, sacrifice and determination. Whether integrated into clandestine operations in occupied territories during wartime or serving as a tool for social change and justice in peacetime, the notion of resistance permeates the fabric of societies and testifies to the indomitable spirit of humankind.

To fully understand resistance, it is essential to examine its contextual nature, recognising that its character can vary depending on the circumstances in which it emerges. By delving into the historical high points of resistance movements, we can dissect the underlying motivations, the challenges encountered, and the integral role these movements have played in the evolution of local and global events.

Understanding context-specific nuances and the dynamic interaction between resistance and authority allows us to appreciate the complex tapestry of human action and empowerment in the face of adversity. By thoroughly exploring the multifaceted nature of resistance, we seek to illuminate the intrinsic link between resistance movements and broader socio-political landscapes, offering deep insight into the depth and enduring relevance of this phenomenon throughout history.

Historical Overview of Resistance Movements

Resistance movements have played an important role throughout history, emerging in various forms to oppose oppressive regimes and occupations. The roots of organised resistance date back to ancient times, with examples such as the Jewish revolt against Roman rule and the numerous uprisings against imperial conquests. However, it was during the 20th century, and more specifically during the Second World War, that resistance movements gained increased attention and recognition due to their decisive influence on the outcome of conflicts. In Europe, the occupied territories saw the emergence of various resistance groups, each with its own motivations, methods, and challenges. These movements embodied the unyielding spirit of individuals and communities determined to defy tyranny and fight for freedom.

The historical backdrop of resistance movements is marked by heroism, sacrifice and an unwavering commitment to resist oppression, often at the risk of one's life. Whether through clandestine operations, acts of sabotage, underground publications, or intelligence gathering, the multifaceted nature of resistance activities reflects the ingenuity and resilience of those involved. To understand these movements, it is essential to consider the complex interaction between local dynamics, societal solidarity, and international support, as well as the complex relationships with occupying forces and collaborating entities. It is essential to recognise the diversity of resistance efforts, which encompass armed rebellions, non-violent protests, clandestine networks, and cultural expressions of dissent.

The historical evolution of resistance movements highlights their ability to adapt to changing challenges and opportunities. While some movements operated in urban centres, others thrived in rural landscapes, using different tactics in their struggle against occupation and totalitarian rule. Each movement had distinct ideological foundations and motivations, ranging from nationalist aspirations to ideological opposition to invaders. The study of historical resistance movements offers valuable insights into human action, resilience, and the pursuit of justice in the face of adversity. Furthermore, it provides a better understanding of the complexities of asymmetric warfare, unconventional strategies, and the lasting impact of grassroots activism on world events. Recognising the historical legacy of resistance movements is a testament to the enduring spirit of humanity and inspires future generations to confront oppression and defend their fundamental rights.

Key figures and leaders

Throughout history, the success of a resistance movement

has been profoundly influenced by the calibre and charisma of its key figures and leaders. These individuals often embody an unwavering determination, moral strength and strategic insight that inspire others to rally behind them in the face of adversity. By examining the essential roles played by these influential personalities, we gain a better understanding of how they shaped the course of resistance movements during times of war and occupation. Key figures and leaders of resistance movements come from a wide variety of backgrounds, ranging from intellectuals and politicians to ordinary citizens driven by fervent patriotism or an unyielding commitment to justice. Among the most famous figures are those who became powerful symbolic icons, such as Jean Moulin in France and Tadeusz Kościuszko in Poland, whose unwavering determination inspired their compatriots to unite against the forces of oppression.

That said, it is essential to recognise the crucial impact of individuals such as Dietrich Bonhoeffer and Sophie Scholl, who challenged the authorities with moral courage and intellectual insight, inspiring active opposition beyond the boundaries of nationality and ideology. These leaders not only directed the practical operations of resistance networks, but also served as moral compasses for the wider population, instilling a sense of purpose and conviction amid the chaos and despair of wartime oppression. The multifaceted leadership styles and approaches adopted by these figures demonstrated their adaptability and resilience. While some excelled at clandestine operations and covert communications, others leveraged their influence to mobilise international support and resources essential to sustaining the resistance effort. Regardless of their specific methods, these leaders shared a common commitment to defending the fundamental values of humanity and justice, determined to counter the encroachments of totalitarianism and enslavement.

The personal sacrifices made by these leaders are emblematic of the broader sacrifices endured by all members of the resistance movement. By enduring imprisonment, torture, and even death, these iconic figures set an extraordinary precedent of selfless devotion, illustrating the principle that the struggle for freedom is worth fighting for at any cost. Their unwavering dedication to the cause continues to inspire future generations, underscoring the lasting impact of their legacy on global narratives of resilience and liberation.

In sum, the indelible mark left by these key figures and leaders of resistance movements testifies to the transformative power of individual action during tumultuous historical periods. Their unwavering determination, strategic foresight, and moral fortitude have bequeathed a lasting legacy that resonates far beyond the confines of their temporal struggles, serving as a source of inspiration and ongoing guidance for those who champion the cause of freedom and justice.

Resistance Strategies and Tactics

The effectiveness of resistance efforts relies on a series of sophisticated strategies and tactics, meticulously crafted to undermine occupying forces and maintain the morale of the oppressed population. One of the fundamental tactics used by resistance movements is clandestine communication, in which networks of couriers, secret codes, and covert transmissions enable the rapid dissemination of essential information while minimising the risk of detection. These communication channels operate within a complex network of trust, established through discreet means and thorough background checks. Another essential aspect of resistance strategy is sabotage and disruption of enemy operations. This can take many forms, from damaging infrastructure and supply lines to infiltrating enemy installations or engaging in acts of economic subversion.

By systematically reducing the occupier's ability to maintain control, resistance movements aim to erode its power and create an environment conducive to liberation. Psychological warfare plays a crucial role in resistance tactics.

Subtle messages, propaganda, and public demonstrations serve to strengthen the spirit of resistance among the population, while sowing doubt and discord among the ranks of the occupying forces. Psychological operations are designed to encourage solidarity and determination, turning public opinion against the oppressors and garnering support for the cause of resistance. In addition to these proactive measures, defensive tactics are equally essential to the survival of resistance movements. Guerrilla warfare, evasion techniques, and the creation of safe havens are all part of a multifaceted defensive strategy aimed at preserving the strength and longevity of the resistance effort. By adapting to the occupier's evolving tactics and demonstrating resilience in the face of adversity, the resistance preserves its ability to continue the fight until it achieves its goals. The adoption of these strategies and tactics is emblematic of the ingenuity and adaptability of resistance movements throughout history. Their effectiveness lies not only in their immediate impact, but also in the lasting inspiration they provide to future generations engaged in similar struggles for freedom and justice.

Technological Contributions to Subversive Efforts

The role of technology in subversive efforts during times of occupation and resistance cannot be underestimated. Throughout history, various innovative technologies have been used to aid resistance movements in their struggle against oppressive regimes. Whether through clandestine communication devices, secret weapons, or tools for sabotage, technology has played a vital role in enabling resistance groups to effectively challenge occupying forces. One of the most significant technological contributions to subversive efforts has been the development and use of clandestine communication devices. These devices have enabled resistance groups to maintain secure and discreet channels for coordinating operations, sharing intelligence, and organising resistance activities, all while evading enemy detection.

Advances in cryptographic technology enabled resistance fighters to securely encode their messages, preventing occupying forces from intercepting and deciphering them. Furthermore, the development and deployment of stealth weapons and sabotage tools significantly enhanced the capabilities of resistance movements. Whether it was the creation of improvised explosive devices (IEDs) or the modification of existing weapons for clandestine operations, technological innovations greatly increased the firepower and effectiveness of resistance groups. In particular, the use of sabotage tools, such as miniaturised explosives and neutralisation devices, has enabled resistance fighters to disrupt enemy infrastructure and supply lines, thereby weakening the occupier's grip on the region.

The use of specialised intelligence-gathering equipment, including hidden cameras, listening devices, and espionage equipment, has provided valuable information about enemy movements and plans, amplifying the strategic advantage of resistance forces. In addition, the adaptation of emerging technologies, such as drones and remote surveillance systems, provided new means of reconnaissance and information gathering, further increasing the operational capacity of resistance groups. Beyond these direct applications, technological advances also had indirect but significant impacts. By leveraging modern technologies, resistance fighters have been able to instil a sense of adaptability and resilience within their movements, demonstrating their ability to evolve in the face of changing tactics employed by occupying forces. The ingenuity of these technological contributions not only facilitated the practical aspects of resistance, but also served to boost the morale of oppressed populations, inspiring hope and determination in the pursuit of liberation. As we delve deeper into the technological contributions to subversive efforts, it becomes clear that the ingenious application of technology played a vital role in empowering resistance movements to face adversity, protect their communities, and ultimately contribute to the collapse of oppressive regimes.

Psychological Impact on Occupied Populations

As the shadow of war enveloped Europe, the psychological

impact on occupied populations became a major concern. Subjugated communities faced a myriad of challenges that tested their resilience and mental fortitude. The experience of living under occupation elicited a complex range of emotions, from fear and despair to defiance and hope. Occupied civilians faced constant anxiety and uncertainty as their daily lives were disrupted by the oppressive presence of the occupying forces. The pervasive atmosphere of fear and surveillance cast a long shadow over the collective psyche, breeding a sense of vulnerability and mistrust.

The imposition of strict regulations and curfews reinforced feelings of confinement and isolation, eroding residents' sense of agency and autonomy. The psychological toll of life under occupation manifested itself in various ways, including increased stress, trauma, and emotional fatigue. That said, the suppression of cultural expression and freedom of speech amplified the population's feelings of alienation and powerlessness. Despite these difficulties, the occupied populations have shown remarkable resilience and courage in the face of adversity. Communities forged bonds of solidarity and mutual support, nurturing a spirit of resistance that transcended the physical constraints of occupation. This indomitable human spirit became a source of inspiration and empowerment, fuelling acts of clandestine defiance and defiance. The psychological impact of occupation has not only shaped the psyche of individuals, it has also left a lasting imprint on the collective memory of nations. Decades after liberation, the legacy of psychological trauma from war continues to reverberate through the narratives and experiences of those who endured occupation. Understanding the psychological repercussions of occupation is essential to grasping the lasting effects of war on the human psyche and highlighting the resilience and tenacity of individuals and communities in the face of adversity.

Collaboration and Conflict: Allies of the Resistance

In times of conflict, the dynamics between resistance movements and allied forces become increasingly complex. The interaction between local resistance groups and allied powers often resulted in a delicate balance between collaboration, support, and, at times, conflicting objectives. While these alliances played a vital role in the fight against occupation, they also gave rise to complex political and military considerations that influenced the outcome of many operations. Relations between resistance fighters and Allied agents were shaped by a myriad of factors, including mutual trust, shared objectives, divergent strategies, and the broader geopolitical landscape. Each side brought unique assets and challenges to the table, fostering an interaction that had a significant impact on the course of resistance efforts. The intertwining of interests sometimes led to scenarios in which the goals of the resistance and the objectives of the Allied powers were not fully aligned, giving rise to potential conflicts and disagreements. Differences in tactics, resource allocation, and long-term goals could create tensions that tested the resilience of collaborative efforts.

Cultural, ideological, and strategic disparities between local resistance fighters and their foreign allies sometimes presented obstacles that required both sides to navigate with skill. Negotiating these complexities demanded a nuanced understanding of each other's perspectives, priorities, and constraints, marking an important chapter in the annals of wartime history. Conversely, collaboration between resistance movements and Allied agents often yielded invaluable intelligence, operational support, and strategic advantage, bolstering the momentum of the resistance and contributing to decisive victories. The exchange of information, resources, and tactical expertise formed the basis for successful joint ventures, amplifying the impact of localised efforts across the entire theatre of war.

Alliances fostered a sense of unity and solidarity that transcended geographical boundaries, inspiring acts of bravery and sacrifice that defined the collective struggle against tyranny. As the war progressed and the tides of battle shifted, the nature of collaboration between resistance groups and the Allies evolved, presenting new challenges and opportunities. Whether through clandestine cooperation, open coordination, or subtle divergence, the complex tapestry of relationships between local resistance movements and Allied powers illuminates a compelling narrative of bravery, cooperation, and the relentless pursuit of freedom. This intricate interaction stands as a testament to the enduring legacy of those who defied oppression and upheaval, shaping the course of history through their unwavering commitment to liberation.

Major operations and their outcomes

During the tumultuous period of the Second World War, several major operations were undertaken by resistance movements in occupied territories, each with its own challenges and objectives. These operations symbolised the daring efforts and unwavering determination of the resistance to thwart the control of the occupying forces and disrupt the status quo. The outcomes of these operations not only left an indelible mark on the course of the war, but also profoundly impacted the annals of history, shaping post-war societies and the geopolitical landscape in ways that continue to influence our world today.

One such notable operation was the Warsaw Uprising, a strategic initiative by the Polish Resistance to liberate Warsaw from Nazi occupation. Despite overwhelming adversity and a lack of outside support, the insurgents fought fiercely for 63 days before succumbing to the superiority of German forces. The massive destruction that followed, coupled with the decimation of the resistance fighters, cast a long shadow over Poland's future, profoundly influencing its national identity and political trajectory.

Operation Anthropoid, a daring mission to assassinate SS-Obergruppenführer Reinhard Heydrich, one of the main architects of the Holocaust, was another major undertaking. This daring act carried out by Czechoslovak resistance agents was punished by the Nazis in the form of the annihilation of the village of Lidice, leaving a lasting scar on both the resistance movement and the local population. However, the successful elimination of Heydrich by the courageous resistance fighters boosted the morale of the oppressed and set off a chain of events that reverberated throughout the war.

Similarly, the French Resistance carried out numerous clandestine operations, including sabotage, intelligence gathering, and direct confrontations with enemy forces, which culminated in the liberation of Paris in August 1944.

The tenacity and sacrifices of the resistance fighters cemented their place in history, reaffirming the belief in the power of unified resistance against oppressive regimes. The aftermath of these monumental operations created a complex tapestry of triumphs, heartbreak, and lasting legacies that continue to shape contemporary perceptions of heroism and defiance in times of war. In the post-war era, the extraordinary feats and tragic losses of these operations have become synonymous with the resilience and sacrifices of those who dared to challenge tyranny, sparking conversations and reflections on the ethical and moral implications of resistance movements. Major operations and their outcomes remain essential to elucidating the multifaceted legacy of resistance, serving as enduring tributes to the unyielding human spirit in the face of adversity.

Long-term effects on post-war societies

The impact of resistance movements on post-war societies has been profound and far-reaching. In the aftermath of the war, many countries found themselves grappling with the lasting effects of occupation, oppression, and conflict. The presence of resistance movements played a vital role in the trajectory of these societies as they sought to rebuild and redefine themselves. One of the most significant long-term effects was the rise of nationalist sentiments and the revitalisation of cultural identity. Resistance movements often embodied and promoted the values and traditions of their respective nations, inspiring a resurgence of national pride and unity. This resurgence helped rebuild the social fabric and foster a sense of collective purpose among citizens of war-torn countries.

Furthermore, the legacy of resistance movements extended to the political landscape, laying the groundwork for democratic reforms and the protection of fundamental human rights. Courageous acts of defiance against authoritarian regimes served as catalysts for the establishment of democratic institutions and the safeguarding of civil liberties. The voices of resistance resonated throughout history, heralding an era of political transformation and societal renewal. Economically, the post-war period saw the reconstruction and rejuvenation of war-ravaged economies. Resistance movements not only contributed to the liberation of territories, but also facilitated the rehabilitation of industries and infrastructure. Their efforts strengthened the foundations for economic recovery and paved the way for sustainable development, fostering resilience in the face of adversity.

The lasting impact of resistance on post-war societies transcended national borders, influencing global discourse on human rights and international relations. The widespread admiration for the indomitable spirit of resistance fighters resonated across continents, inspiring a renewed commitment to universal principles of justice and freedom. The lessons learned from the struggles of resistance movements continue to fuel contemporary debates on humanitarian intervention, diplomacy, and the responsibilities of the international community.

In short, the profound implications of resistance movements on post-war societies testify to the resilience and fortitude of individuals and communities in their quest for freedom and justice. Their enduring legacy is a poignant reminder of the lasting power of unity, courage and determination to shape the course of history.

Final Analysis: Lessons for Modern Times

When we reflect on the enduring legacy of resistance movements during and after the war, it becomes clear that these historic efforts offer invaluable lessons for modern societies. The resistance, courage, and unwavering determination demonstrated by the individuals and groups involved in resistance efforts are timeless examples of the human spirit's ability to overcome adversity. It is imperative to recognise the enduring relevance of these experiences and draw meaningful parallels with contemporary challenges. One of the most striking lessons for modern times is the power of unity and solidarity in the face of oppression. The coordinated actions of diverse individuals within resistance movements demonstrated the potential of collective action to effect substantial change, even in the most dire circumstances. This underscores the importance of fostering inclusive and collaborative communities capable of confronting systemic injustices and totalitarian regimes.

The strategic ingenuity displayed by resistance fighters in subverting oppressive forces offers crucial insights for confronting current threats. Their ability to adapt, innovate, and employ unconventional tactics demonstrates the need for creativity and flexibility in addressing complex geopolitical and social challenges.

The fortitude and moral compass demonstrated by those who risked their lives to resist tyranny inspire introspection in an era marked by ethical dilemmas and global crises. The bravery of individuals who adhered to universal principles of justice and righteousness highlights the enduring importance of maintaining moral integrity in the face of adversity.

The unyielding pursuit of truth and transparency in a climate of deception and misinformation is a poignant reminder of the essential role of truthfulness in safeguarding democratic values and human rights. These historical lessons compel us to critically examine the current dynamics of our society and engage in constructive dialogue to resolve contemporary moral and ethical dilemmas. Finally, the long-term impact of resistance efforts on post-war societies offers valuable insights into the complexity of rebuilding fractured communities and reconciling historical grievances. By studying the successes and failures of post-conflict reconstruction, modern societies can learn essential lessons for fostering lasting peace, reconciliation, and societal healing in the aftermath of contemporary conflicts. As we face the challenges of the 21st century, the enduring legacy of resistance movements testifies to the indomitable spirit of humanity and provides essential guidelines for addressing the multifaceted challenges of our time.

18
Reflections
Ethical Implications and the Future of Innovation

Ethical considerations

To explore the complex interaction between technology and ethics, it is essential to take a journey back in time, delving into the historical roots of the ethical dilemmas arising from technological advances. The evolution of ethical considerations in the field of technology reveals a fascinating narrative marked by pivotal moments that have shaped societal perceptions and moral frameworks. From the Industrial Revolution to the Digital Age, technological innovation has necessitated an ongoing discourse on the ethical implications of progress. This transhistorical exploration not only highlights the complexities inherent in navigating ethical landscapes, but also provides valuable insights into the dynamic nature of ethical reflections. As the annals of history reveal, instances of ethical violations precipitated by technological prowess are salient reminders of the profound impact of human ingenuity on moral dilemmas. Thus, this retrospective odyssey offers a perspective for understanding the interconnectedness of past ethical transgressions and contemporary ethical considerations. By examining the ethical issues posed by previous technological breakthroughs, we discern recurring themes that transcend temporal boundaries and underscore the enduring relevance of ethical reflections. By immersing ourselves in this historical tapestry, we cultivate a keen sense of contextual richness, enabling us to assess current ethical dilemmas in a more nuanced manner. Consequently, the following analysis of historical ethical lapses aims to foster a comprehensive understanding of the complex interaction between technology and morality, thereby setting the stage for an incisive examination of contemporary ethical conundrums.

Historical Analysis of Ethical Breaches

Throughout history, ethical breaches have been common in the realm of technological and scientific advances. These breaches often result from the intersection of innovation, power, and moral decision-making, with repercussions that ripple across generations. One of the earliest examples of ethical breaches dates back to nuclear technology in the 20th century. The development and use of atomic bombs during World War II raised profound ethical concerns. The resulting devastation and long-term consequences prompted global introspection on the ethical implications of harnessing such immense power.

The controversial Tuskegee syphilis study in the United States is a poignant example of medical research that went ethically awry. This infamous study, conducted over several decades, prevented African American participants from receiving appropriate treatment to observe the natural progression of syphilis, without their informed consent. It illustrates a flagrant violation of human rights, highlighting the critical need for ethical guidelines in scientific research. In the modern era, the advent of the internet and rapid digitisation has also given rise to ethical dilemmas. Issues such as privacy breaches, data exploitation, and cyber warfare have highlighted the need for ethical considerations in the field of technology. In particular, the Cambridge Analytica scandal laid bare the ethical grey areas surrounding the collection and manipulation of data for political purposes. This historical retrospective serves as a poignant reminder of the considerable impact that ethical lapses can have on individuals, societies and global stability. Understanding these past transgressions is essential to shaping our approach to innovation and ensuring that ethical values remain at the heart of technological and scientific progress.

Technology and moral responsibility

Technological advances often bring about profound changes in society, revolutionising the way we live, work and interact. However, these transformative innovations raise a crucial question: that of moral responsibility. As the power and reach of technology expands, it becomes imperative to reflect on the ethical implications of its use. The intersection of technology and moral responsibility is a complex landscape that requires careful navigation and consideration. At the heart of this discussion is the fundamental question of how technological progress should be ethically governed and harnessed for the collective good. In the realm of emerging technologies such as artificial intelligence, biotechnology, and data privacy, ethical dilemmas abound. The development and deployment of artificial intelligence systems raise concerns about bias, accountability, and the potential erosion of human autonomy. Similarly, advances in biotechnology pose ethical questions regarding genetic manipulation, the commodification of life, and equitable access to healthcare innovations.

The ubiquitous collection and use of personal data in the digital age raises questions about privacy, consent, and the responsible management of sensitive information. These challenges underscore the imperative of moral responsibility in technological innovation.

The influence of technology on societal norms cannot be underestimated. Innovations have the power to shape cultural values, redefine interpersonal relationships and restructure economic systems. The advent of social media platforms, for example, has changed communication patterns, reshaped public discourse and given rise to new forms of social interaction. However, this transformation has also created ethical dilemmas related to online harassment, misinformation, and the erosion of privacy boundaries. The technology industry therefore bears a profound moral responsibility to consider the broader societal implications of its innovations. As technological advances continue to accelerate, stakeholders across different sectors must actively engage in dialogues centred on ethical considerations. This involves fostering interdisciplinary collaborations between technologists, ethicists, policymakers, and the public to ensure that the trajectory of technological progress aligns with moral imperatives.

Organisations must cultivate cultures that prioritise ethical decision-making, transparency, and accountability in the design and implementation of technological solutions. By promoting a strong ethic of moral responsibility, the transformative potential of technology can be realised while protecting against its unintended negative consequences.

The role of innovation in shaping societal norms

In the realm of technological progress, innovation is not only a catalyst for progress, but also a force that shapes societal norms. Throughout history, transformative innovations have consistently reshaped the way individuals and communities interact, communicate, and conduct their daily lives. From the printing press to the Industrial Revolution to the digital age, every significant technological advance has left an indelible mark on society's values, behaviours and expectations. Innovations, especially those that are adopted en masse, have the power to redefine social structures and dynamics. For example, the widespread use of social media platforms has fundamentally changed the way people connect, share information and form opinions. The influence of these platforms on public discourse, political engagement and cultural movements reflects the profound impact of technological innovation on shaping societal norms.

Innovation often fosters the emergence of new industries and professions, leading to changes in professional landscapes and skill requirements. As automation and AI continue to revolutionise industries, the nature of work and employment is undergoing a significant transformation, thereby redefining societal perceptions of career paths, economic stability, and social mobility.

Ethical considerations surrounding emerging technologies play a central role in influencing societal norms. Controversies surrounding privacy, data security, and the ethics of artificial intelligence shape public perception and awareness, sparking conversations and advocacy around regulatory frameworks and responsible innovation practices. How these ethical dilemmas are addressed and integrated into technological development can shape the moral compass of future generations.

Technology has the capacity to amplify cultural and ideological shifts, contributing to the redefinition of moral and ethical boundaries. Global interconnectedness facilitated by technological innovations exposes individuals to diverse perspectives and worldviews, sparking discussions and debates about tolerance, inclusion, and human rights. This interaction between innovation and societal norms clearly highlights the complex relationship between technological progress and its impact on the formation of collective values.

In short, the role of innovation in shaping societal norms encompasses a multifaceted and dynamic process that goes beyond its technological implications. Its profound influence on communication, employment, ethics, and cultural evolution highlights the considerable impact of innovation on the social fabric. At each significant stage of technological progress, it becomes increasingly vital to critically assess the ramifications of innovation on societal norms and to uphold ethical principles while navigating this transformative landscape.

Lessons from the legacy of resistance

The legacy of wartime resistance offers valuable lessons for contemporary society. Throughout history, we discover compelling stories of individuals and organisations who challenged authoritarian regimes, risking their lives to defend moral and ethical principles. The courage displayed by these unsung heroes resonates through time, serving as a poignant reminder of the enduring power of resilience and unwavering dedication to principles of justice and humanity. By immersing ourselves in these stories, we uncover valuable insights that strike a deep chord in the realm of technological innovation and ethical responsibility. These lessons not only allow us to understand the monumental sacrifices made by those who resisted tyranny, but also serve as a catalyst for shaping our ethical compass in the sphere of modern technology. By carefully examining acts of historical defiance, we draw parallels with the current landscape of technological progress. The courage displayed by individuals who challenged oppressive systems offers a stark contrast to the challenges posed today by ethical dilemmas surrounding technological development.

The principles underlying acts of resistance, such as integrity, empathy, and unwavering dedication to the common good, serve as beacons for today's innovators and technologists. Their courageous actions instil in us a commitment to uphold ethical standards and assume moral responsibility in the relentless pursuit of progress. It is imperative that we recognise and honour stories of resistance, for they contain invaluable wisdom that transcends time and testifies to the indomitable spirit of human perseverance.

Examining the ethics of resistance allows us to deeply appreciate the essential role that ethical decision-making plays in the trajectory of technological progress. These invaluable lessons compel us to continually re-evaluate our ethical frameworks and cultivate a culture of accountability, transparency, and ethical rigour. Therefore, the legacy of resistance is a link between the past and the present, inspiring us to weave the virtues of integrity, resilience, and moral fortitude into the fabric of technological innovation, ensuring that the legacy of ethical responsibility endures for generations to come.

Contemporary Challenges in Technological Ethics

As technological advances continue to reshape our world, the ethical implications of these innovations are becoming increasingly complex and multifaceted. From artificial intelligence to data privacy, genetic engineering to cybersecurity, contemporary society faces a myriad of challenges in navigating the changing landscape of technology ethics. One of the key contemporary challenges is the ethical use of AI, which raises concerns about bias, discrimination, and the potential loss of human control over decision-making processes.

The collection, storage, and use of large amounts of personal data have sparked discussions about invasions of privacy and individual freedoms. In addition, the rapidly expanding field of genetic engineering presents ethical dilemmas related to the manipulation of human genomes and the risk of unintended consequences. Cybersecurity is also emerging as a key concern, given the growing prevalence of cyber threats and their potential impact on individuals, organisations and even nations. In the face of these challenges, it is essential to ensure global compliance and regulatory frameworks that govern the responsible use of technology. The ever-expanding nature of technology requires a consistent approach to ethical guidelines and international regulations, harmonising different perspectives and standards to ensure universal adherence to ethical practices.

Promoting an ethical culture within technological development requires a concerted effort to instil values of integrity, transparency and accountability throughout the industry. Addressing contemporary challenges in technology ethics also involves anticipating future technological innovations and their potential ethical implications. This proactive approach requires ongoing dialogue and collaboration between stakeholders, policymakers, and technical experts to anticipate and address potential ethical obstacles before they arise. Ultimately, addressing contemporary challenges in technology ethics requires a holistic and forward-looking approach that transcends national borders and industrial sectors, placing ethical considerations at the forefront of technological progress.

Regulatory frameworks and global compliance

In our interconnected world, the ethical implications of technological innovation transcend national borders. Regulatory frameworks play a crucial role in the responsible development and deployment of advanced technologies. As advances in science and engineering continue to accelerate, it is imperative that nations collaborate to establish global compliance standards that prioritise ethical considerations. The proliferation of advanced technologies, such as artificial intelligence, biotechnology, and autonomous systems, requires international consensus on regulatory principles to guard against potential abuse and harm.

One of the key challenges in formulating global regulatory frameworks is striking a delicate balance between promoting innovation and upholding ethics. Achieving this balance requires policymakers, business leaders, ethicists, and international organisations to collaborate in developing comprehensive guidelines that foster technological progress while respecting universal moral principles. Furthermore, it is essential to consider disparities among nations in terms of technological capabilities and resources to prevent disproportionate ethical violations and promote equity in access to and use of cutting-edge innovations. Establishing robust global compliance mechanisms requires a multifaceted approach. This includes harmonising legal frameworks, standardising ethical protocols, and creating oversight bodies with the power to monitor and enforce ethical guidelines.

Promoting transparency and accountability within the technology sector is essential to cultivating a culture of ethical conduct and responsible innovation on a global scale. To this end, intergovernmental agreements, multinational initiatives, and bilateral collaborations play a vital role in developing a coherent regulatory landscape that transcends geopolitical boundaries. Adherence to global compliance standards strengthens ethical safeguards and builds trust in emerging technologies. It inspires confidence among stakeholders, including governments, businesses, and the public, by demonstrating a commitment to upholding ethical standards in the development and deployment of technologies. This, in turn, facilitates the integration of innovative solutions across various areas of society, from healthcare and governance to environmental conservation and economic advancement. However, navigating the complexities of global compliance requires constant reflection, refinement and adaptation of rules to keep pace with the dynamic nature of technological change. Periodic reassessment of regulatory frameworks and their alignment with ethical imperatives is essential to effectively address emerging ethical challenges. Furthermore, promoting a culture of awareness, education, and ethical responsibility, both at the individual and organisational levels, is fundamental to ensuring sustained adherence to global compliance standards.

At the dawn of unprecedented technological advances, the formulation and implementation of robust global compliance frameworks are the cornerstone of building a future where innovation thrives in accordance with ethical values. By encouraging international cooperation, knowledge sharing, and mutual accountability, nations can collectively steer the trajectory of technological progress toward a future defined by principled innovation and global well-being.

Fostering an ethical culture in technological development

In the field of technological development, promoting an ethical culture is essential to ensuring that innovation aligns with societal values and moral principles. This involves establishing comprehensive ethical guidelines and a commitment to integrity at every stage of technological development. To cultivate such a culture, organisations must prioritise ethical considerations in decision-making processes, from design to implementation. A crucial aspect is promoting transparency and accountability, where stakeholders are held responsible for the ethical implications of their contributions to technological innovations.

It is essential to create an environment that encourages open discourse on ethical dilemmas and implications, as this allows diverse perspectives to shape the ethical trajectory of technological development. Promoting an ethical culture also requires ongoing education and training initiatives to equip professionals with the knowledge and skills necessary to address the complex ethical challenges inherent in technological progress.

Integrating ethical assessments into the heart of development processes not only strengthens the moral compass of projects, but also fosters a collective sense of responsibility within teams. Embracing diversity and inclusivity in the technological landscape is fundamental to shaping an ethical culture, as varied perspectives help uncover potential ethical blind spots and improve the overall framework for ethical decision-making. Furthermore, establishing partnerships and collaborations with ethicists, academics, and regulatory bodies can offer valuable insights and guidance for navigating the complex intersection of technology and morality. Ultimately, promoting an ethical culture in technological development provides proactive protection against the unintended perpetuation of harmful consequences and ensures that innovation remains aligned with the greater good of society.

Forecasting the Future: Technology and Morality

Rapid advances in technology have consistently raised complex ethical dilemmas that require careful consideration. As we enter an era marked by artificial intelligence, quantum computing, and biotechnological breakthroughs, the intersection of technology and morality becomes increasingly complex. The ability of innovation to both reinforce and compromise human values and societal norms has prompted critical examination of the future landscape. Predicting the trajectory of technology's impact on ethics requires a multifaceted approach that takes various considerations into account. One key area of concern relates to the potential ramifications of profound technological disruption on established moral frameworks. As we delve deeper into areas such as genetic engineering, augmented reality, and autonomous systems, it becomes imperative to anticipate how these advances will intersect with fundamental ethical principles. The need to assess potential implications for individual autonomy, privacy, and fairness requires constant vigilance and robust ethical guidelines.

The evolution of the fabric of social relationships and interactions in an increasingly digital world is a crucial area for ethical foresight. The prevalence of virtual environments, online communication and data-driven decision-making requires careful assessment of the ethical dimensions inherent in these contexts. Issues of digital rights, algorithmic bias, and the erosion of privacy underscore the imperative to proactively shape the ethical foundations of our technology-dependent society. Anticipating potential ethical issues also requires exploring the implications of technology for work, economic disparities, and global cooperation. As automation and artificial intelligence continue to transform industries and labour markets, ethical considerations related to unemployment, reskilling, and the equitable distribution of technological benefits are at the forefront.

The ethical dimensions of cross-border data governance, technology diplomacy, and security imperatives demand sustained attention in the pursuit of a harmonised and ethical technology ecosystem. Despite the complexities and uncertainties surrounding the relationship between technology and morality, several guiding principles can help us anticipate the challenges ahead. Prioritising transparency, accountability, and inclusion in technological development is essential to fostering an ethical foundation. Promoting interdisciplinary dialogue, cultivating technological literacy, and involving diverse voices in ethical deliberations are essential strategies for navigating the evolving terrain of technology and morality. By engaging in proactive discourse and strategic foresight, we can strive to shape a future where technological progress aligns harmoniously with sustainable moral imperatives.

Conclusion: Moving forward with ethical integrity

As we conclude our exploration of the ethical implications and future of innovation, it becomes clear that a genuine commitment to ethical integrity is essential to guiding technological progress. The rapid advancement of technology demands a proactive approach to addressing ethical challenges and ensuring responsible innovation. Moving forward, it is imperative that individuals, businesses, and governments prioritise ethical considerations in all facets of technological development. Maintaining ethical integrity requires a comprehensive understanding of the potential consequences of each innovation on the broader societal landscape. Embracing ethical integrity also means learning from historical precedents, where ethical lapses have led to devastating outcomes. By acknowledging past mistakes and implementing strict ethical standards, we can pave the way for a future where technological progress aligns harmoniously with moral responsibilities. This journey toward ethical integrity begins with promoting a culture of conscientious innovation, where ethical considerations are integrated into every stage of technological development. From initial conceptualisation to implementation and beyond, ethical assessments must be continually undertaken as a fundamental part of the innovation process.

The role of regulatory frameworks in maintaining ethical integrity cannot be overstated. Governments and international organisations must collaborate to establish and enforce regulations that govern the ethical use of technologies. These frameworks provide a vital structure for promoting responsible innovation and ensuring that ethical boundaries are respected in global technological landscapes. Furthermore, it is essential to foster transparency and accountability within the technology sector. Honest and open communication about the ethical implications of innovations promotes trust among stakeholders and the wider community. The future of innovation rests on the principle of aligning technology with ethical principles. Anticipating the ethical implications of emerging technologies and futuristic advances is essential to proactively addressing potential challenges. It is essential to engage in thoughtful discourse on the societal impact of innovation, emphasising the ethical considerations that guide the trajectory of technological evolution. In doing so, we can collectively steer the course of innovation towards a future characterised by ethical integrity and moral responsibility.

At its core, moving forward with ethical integrity requires an unwavering dedication to prioritising the well-being of humanity above all else. This necessitates a paradigm shift where ethical considerations are seamlessly woven into the fabric of technological progress. With ethical integrity as our compass, we embark on a future where innovation will be aligned with our collective values and guided by a deep sense of ethical responsibility.

References
For Further Reading

- BARBIER, M. K., & SHOWALTER, D. (2017). Operation Paperclip—Antecedents and Dubious Draftees. In Spies, Lies, and Citizenship: The Hunt for Nazi Criminals (pp. 211–232). University of Nebraska Press. https://doi.org/10.2307/j.ctt1tqx72k.16

- BARNWELL, M. (2021). MODERNIZATION 1919–1967: Destruction, Reconstruction, New World Order. In Design and Culture: A Transdisciplinary History (pp. 111–162). Purdue University Press. https://doi.org/10.2307/j.ctv15pjxxx.6

- Baker, M. E., & Hughes, K. (1991). Fifty Years of Excellence: The Redstone Arsenal Complex Since 1941. Army History, 20, 25–30. http://www.jstor.org/stable/26302854

-Bennett, C. (2024). Roswell Revealed: The Untold Story Of America's Most Famous UFO Incident. Global East-West (London).

- Beyerchen, A. (1982). German Scientists and Research

Institutions in Allied Occupation Policy. History of Education Quarterly, 22(3), 289–299. https://doi.org/10.2307/367770

- Beyerchen, A. (1992). German Imports [Review of Secret Agenda. The United States Government, Nazi Scientists, and Project Paperclip, 1945 to 1990., by L. Hunt]. Science, 255(5043), 481–482. http://www.jstor.org/stable/2876027

- BIBES, G. (1968). LE FASCISME ITALIEN: ÉTAT DES TRAVAUX DEPUIS 1945. Revue Française de Science Politique, 18(6), 1191–1244. http://www.jstor.org/stable/43115245

- Brustein, W. (1991). The "Red Menace" and the Rise of Italian Fascism. American Sociological Review, 56(5), 652–664. https://doi.org/100.2307/2096086

- Capitan, C. (1963). FASCISME ET FASCISMES. Cahiers Internationaux de Sociologie, 35, 165–175. http://www.jstor.org/stable/40689246

- DOERRIES, R. R. (2006). Recently Declassified American Records of the Third Reich and Its Aftermath [Review of U.S. Intelligence and the Nazis, by R. Breitman, N. J. W. Goda, T. Naftali, & R. Wolfe]. Diplomatic History, 30(2), 301–305. http://www.jstor.org/stable/24915097

- Farquharson, J. (1997). Governed or Exploited? The British Acquisition of German Technology, 1945-48. Journal of Contemporary History, 32(1), 23–42. http://www.jstor.org/stable/261074

- Front Matter. (2015). Air Power History, 62(1). https://w

ww.jstor.org/stable/26276559

- Fujitani, T. (2007). Right to Kill, Right to Make Live: Koreans as Japanese and Japanese as Americans During WWII. Representations, 99(1), 13–39. https://doi.org/10.1525/rep.2007.99.1.13

- GIMBEL, J. (1986). U.S. Policy and German Scientists: The Early Cold War. Political Science Quarterly, 101(3), 433–451. https://doi.org/10.2307/2151624

- GIMBEL, J. (1990). German Scientists, United States Denazification Policy, and the "Paperclip Conspiracy." The International History Review, 12(3), 441–465. http://www.jstor.org/stable/40106226

- GIMBEL, J. (1990). Project Paperclip: German Scientists, American Policy, and the Cold War. Diplomatic History, 14(3), 343–365. http://www.jstor.org/stable/24911848

- GIMBEL, J. (1990). The American Exploitation of German Technical Know-How after World War II. Political Science Quarterly, 105(2), 295–309. https://doi.org/10.2307/2151027

- Gilbert, M. (2002). The Rise of Fascism in Europe in the twentieth century: lessons for today. India International Centre Quarterly, 29(2), 31–38. http://www.jstor.org/stable/23005773

- Grunden, W. E., Kawamura, Y., Kolchinsky, E., Maier, H., & Yamazaki, M. (2005). Laying the Foundation for Wartime Research: A Comparative Overview of Science Mobilization in National Socialist Germany, Japan, and the Soviet Union. Osiris, 20, 79–106. http://www.jstor.org/stable/3655252

- Hassner, P. (1991). L'Europe et le spectre des nationalismes. Esprit (1940-), 175 (10), 5–22. http://www.jstor.org/stable/24274754

- Heggie, V. (2014). Why Isn't Exploration a Science? Isis, 105(2), 318–334. https://doi.org/10.1086/676569

- HIGGINS, A. S. (2023). The Ghost of Sputnik. In Higher Education for All: Racial Inequality, Cold War Liberalism, and the California Master Plan (pp. 9–36). University of North Carolina Press. http://www.jstor.org/stable/10.5149/9781469672939_higgins.5

- Huddleston, R., Jacobsen, A., & Lichtblau, E. (2015). [Review of Operation Paperclip: The Secret Intelligence Program that Brought Nazi Scientists to America; The Nazis Next Door: How America Became a Safe Haven for Hitler's Men]. Air Power History, 62(1), 52–53. https://www.jstor.org/stable/26276571

- Jones, E. (2006). "LMF": The Use of Psychiatric Stigma in the Royal Air Force during the Second World War. The Journal of Military History, 70(2), 439–458. http://www.jstor.org/stable/4137960

- KASPAREK, C. (2002). ENIGMA AND POLAND REVISITED. The Polish Review, 47(1), 97–103. http://www.jstor.org/stable/25779307

- Kogan, N. (1969). The Origins of Italian Fascism. Polity, 2(1), 100–105. https://doi.org/10.2307/3234092

- KURLANDER, E. (2017). NAZI TWILIGHT: Miracle Weapons, Supernatural Partisans, and the Collapse of the Third Reich. In Hitler's Monsters (pp. 263–296). Yale University Press. http://www.jstor.org/stable/j.ctt1q31shs.14

- Kuzmarov, J. (2020). "There's Something Rotten in Denmark": Frank Olson and the Macabre Fate of a CIA Whistleblower in the Early Cold War. Class, Race and Corporate Power, 8(1). https://www.jstor.org/stable/48645495

- LANEY, M. (2019). Setting the Stage to Bring in the "Highly Skilled": Project Paperclip and the Recruitment of German Specialists after World War II. In M. MARINARI, M. Y. HSU, & M. C. GARCÍA (Eds.), A Nation of Immigrants Reconsidered: US Society in an Age of Restriction, 1924-1965 (pp. 144–160). University of Illinois Press. https://doi.org/10.5406/j.ctv9b2wjb.13

- Lambeth, B. S. (2003). THE AIR FORCE'S STRUGGLE FOR SPACE. In Mastering the Ultimate High Ground: Next Steps in the Military Uses of Space (pp. 9–36). RAND Corporation. http://www.jstor.org/stable/10.7249/mr1649af.8

- Lindsay, J. R. (2025). Espionage: Bletchley Park and the Mechanization of Intelligence. In Age of Deception: Cybersecurity as Secret Statecraft (pp. 97–124). Cornell University Press. http://www.jstor.org/stable/10.7591/jj.24033720.8

- Lohaus, P. (2014). US Conventional and Special Operations Forces since World War II. In A Precarious Balance: Preserving the Right Mix of Conventional and Special Operations Forces (pp. 3–30). American Enterprise Institute. http://www.jstor.org/stable/resrep03192.5

- Loewenstein, K. (1937). Dictatorship and the German Constitution: 1933-1937. The University of Chicago Law Review, 4(4), 537–574. https://doi.org/10.2307/1596654

- Lundquist, Charles A. (2014). Transplanted Rocket Pioneers. [Documents]. The University of Alabama in Huntsville. https://jstor.org/stable/community.34308176

- O'Reagan, D. M. (2015). French Scientific Exploitation and Technology Transfer from Germany in the Diplomatic Context of the Early Cold War. The International History Review, 37(2), 366–385. http://www.jstor.org/stable/24703240

- Paxton, R. O. (1995). Les fascismes essai d'histoire comparée. Vingtième Siècle. Revue d'histoire, 45, 3–13. https://doi.org/10.2307/3771012

- PIANTADOSI, C. A. (2012). THE EXPLORERS. In Mankind Beyond Earth: The History, Science, and Future of Human Space Exploration (pp. 45–67). Columbia University Press. http://www.jstor.org/stable/10.7312/pian16242.7

- Price, D. H. (2011). HOW THE CIA AND PENTAGON HARNESSED ANTHROPOLOGICAL RESEARCH DURING THE SECOND WORLD WAR AND COLD WAR WITH LITTLE CRITICAL NOTICE. Journal of Anthropological Research, 67(3),

333-356. http://www.jstor.org/stable/41303322

- Sand, S. (1983). L'IDÉOLOGIE FASCISTE EN FRANCE. Esprit (1940-), 80/81 (8/9), 149-160. http://www.jstor.org/stable/24270217

- Seed, D. (2009). [Review of Doomsday Men: The Real Dr. Strangelove and the Dream of the Superweapon, by P. D. Smith]. The Modern Language Review, 104(1), 195-196. http://www.jstor.org/stable/20468174

- Sheehan, W., & Bell, J. (2021). Marsniks and Flyby Mariners: The 1960s. In Discovering Mars: A History of Observation and Exploration of the Red Planet (pp. 231-259). University of Arizona Press. https://doi.org/10.2307/j.ctv1zqdv0b.15

- Sherman, W. H. (2021). ENCRYPTING/DECRYPTING. In A. Blair, P. Duguid, A.-S. Goeing, & A. Grafton (Eds.), Information: A Historical Companion (pp. 417-423). Princeton University Press. https://doi.org/10.2307/j.ctv1pdrrbs.46

- Siddiqi, A. A. (2004). Russians in Germany: Founding the Post-War Missile Programme. Europe-Asia Studies, 56(8), 1131-1156. http://www.jstor.org/stable/4147400

- Siddiqi, A. A. (2009). Germans in Russia: Cold War, Technology Transfer, and National Identity. Osiris, 24(1), 120-143. https://doi.org/10.1086/605972

- Smith, M. G. (2023). SATELLITE VISIONS: The Dilemmas of Space Exploration. In The Rocket Lab: Maurice Zucrow,

Purdue University, and America's Race to Space (pp. 154–178). Purdue University Press. https://doi.org/10.2307/j.ctv2x1nrw5.11

- STAHNISCH, F. W. (2020). The Machtergreifung of the National Socialists and Its Effects on the German-Speaking Neurosciences: Marginalization – Oppression – Forced Migration. In A New Field in Mind: A History of Interdisciplinarity in the Early Brain Sciences (Vol. 52, pp. 201–236). McGill-Queen's University Press. https://doi.org/10.2307/j.ctv10kmfd5.12

- Țurcanu, F. (2007). Une guerre oubliée: la Première Guerre mondiale. Cités, 29, 157–160. http://www.jstor.org/stable/40621392

- Weisbrod, B. (2003). The Moratorium of the Mandarins and the Self-Denazification of German Academe: A View from Göttingen. Contemporary European History, 12(1), 47–69. http://www.jstor.org/stable/20081140

- Werth, K. (2004). A Surrogate for War—The U.S. Space Program in the 1960s. Amerikastudien / American Studies, 49(4), 563–587. http://www.jstor.org/stable/41158096

- Wey, A. L. K. (2017). Special Operations by Air Power: Strategic lessons from World War II. Air Power History, 64(1), 33–40. https://www.jstor.org/stable/26276840

www.ingramcontent.com/pod-product-compliance
Lightning Source LLC
Chambersburg PA
CBHW071229070526
44583CB00017B/2111